The Newcom++ Vision Book

Sergio Benedetto · Luis M. Correia · Marco Luise
Editors

The Newcom++ Vision Book

Perspectives of Research on Wireless Communications in Europe

Springer

Sergio Benedetto
ISMB, Istituto Superiore Mario Boella
Torino, Italy
NEWCOM++ Scientific Director

Marco Luise
CNIT, Consorzio Nazionale Interuniversitario
per le Telecomunicazioni
Pisa, Italy
NEWCOM++ Managing Director

Luis M. Correia
Instituto Superior Técnico / Instituto de Telecomunicações
University of Lisbon, Portugal
Head of NEWCOM++ Integration Activites

ISBN 978-88-470-1982-9 ISBN 978-88-470-1983-6 (eBook)
DOI 10.1007/978-88-470-1983-6

Library of Congress Control Number: 2011927611

Springer Milan Heidelberg New York Dordrecht London

Cover design: Beatrice 8, Milan

Typesetting with LaTeX: PTP-Berlin, Protago TeX-Production GmbH, Germany (www.ptp-berlin.eu)

Springer-Verlag Italia S.r.l., Via Decembrio 28, I-20137 Milano
Springer is a part of Springer Science+Business Media (www.springer.com)

Foreword

This book is about visions. Every sphere of human undertakings needs visions to proceed and to succeed. So do we, the engineering-scientist community in wireless mobile communications. However, early visions arose from sources other than engineering. As early as 1932, just after radio broadcasting had had a fledgling start and had been newly discovered as a means of propaganda, the German writer Erich Kästner had a vision that he expressed in a children's novel, *The 35th of May, or Conrad's Ride to the South Seas*: "But most impressive was this: A gentleman, who was coasting in front of them on the moving sidewalk, stepped down on the street, took a telephone from his pocket, spoke a number into the microphone and said: "Gertrud, listen, I'll be an hour late for dinner tonight. I have to go to the lab first. See you, honey!" Then he put away his pocket phone, stepped onto the moving belt again, continued reading his book and off he went." What a vision! The only thing Kästner did not foresee was that we would now telephone while on the move. It is all but a coincidence that a man of letters would voice such a prophecy. Language and communication are constitutive elements of mankind. We define ourselves by our social interactions, and for such we need to speak to each other. Some philosophers go so far to say that speaking to each other makes us human. Wireless mobile communications turned out to be the most rapid technology ever to reach a billion people (taking just 17 years) because it fulfils a basic human need: communication. The changes that wireless communications have brought about in society are enormous. It was not just becoming free from a certain location, the termination of a phone line, to be able to communicate over distance. It is the way we interact today socially, how we plan our activities, both private and business, how businesses have been accelerated by push-email. All these changes are unprecedented, as are mobile micro payments that have enabled economic growth in regions where banks are not accessible, but also the blurring of work and leisure times.

Societal issues and needs are the true drivers of technology. Thus, I welcome and applaud the concept of the *NEWCOM++ Vision Book* starting from the viewpoint of society's needs.

It is debatable whether wireless communication was invented in Europe, beginning with Guglielmo Marconi's pioneering work. In terms of *technology* alone, researchers from Motorola and AT&T' Bell Labs, claiming priority for new ideas in telecommunications (as usual), and Japan's NTT have set undeniable, significant milestones. But, at the 1984 news conference announcing the divestiture of AT&T, the outgoing CEO, Charles Lee Brown, was asked what would become of AT&T's infant cellular properties. He could not give an answer. Either he wasn't aware there was such a business or, at least, he wasn't aware of its potential. In terms of *systems*, on the other hand, GSM ended the frustratingly incompatible standards for public wireless communications in operation in the early 1980's in Europe, and eventually in the world. Remember that GSM is rooted in a political decision taken by visionary European politicians at that time. Since then, European *industry* and *academia* have accomplished a magnificent job originating from the political vision of a pan-European mobile telephone system to the current de facto global standard. After the Paris 1986 technology beauty contest, a major decision was taken for a narrowband TDMA access method, following essentially Scandinavian innovations. Interestingly enough, at that time, going digital was not an obvious decision but a courageous one. The European Commission clearly and deliberately saw GSM and its follower technologies (UMTS, HSPA, LTE) as a means to create a mass market and started to fund R&D in wireless communications. The funding comprised a two-tier approach, a cheap one and an expensive one. The expensive one, let me call it the "D" branch of R&D, started out modestly with a single Mobile Communications project in RACE I (i.e. the legendary R 1043), but soon exploded to innumerable projects. Making up the majority of projects in the Third, the Fourth, the Fifth, and the Sixth Framework Programmes of European funded industrial R&D, they however provided ever diminishing returns. After the enthusiastic, truly innovative period had faded out, some lines of research turned into run-on projects without vision. Most of the money flowed into large companies, some of which ceased to exist, while underfunded universities were condescendingly included in the consortia in order to have at least some novel ideas to present at the end of the project.

The cheap path, let us identify it as the "R" branch, started with COST Action 207 *Digital Land Mobile Communications* (COST stands for European Cooperation in the field of Scientific and Technical research) and was, beyond comparison, cheaper in terms of budget spent by the Commission. With participation of public and private telecom research centers, four universities and the research branches of several European telecom operators – indeed, in the days of monopoly operators, there did exist dedicated research branches, some even staffed with thousands of researchers and engineers! – clarified many open technological issues in GSM's specification phase. A key contributor to GSM's success was the cooperation among colleagues modeling the radio channel and those who would use these models to base their decisions on competing proposed technological solutions. The agreed-on "typical models" – I would prefer the term "reference models" – for rural areas, (RA), typical urban (TU), bad urban (BU) and hilly terrain (HT) have become legendary. The successor actions, COST 231, 259, 273, and 2100 turned out to be as high-yield for wireless as low-budget for the Commission.

The general discontent with the situation of heavily funded industrial R&D and underfunded academic research in many technology areas finally lead to the establishment of Networks of Excellence in European research, of which NEWCOM++ is a grand example. NEWCOM++ was graded as the best proposal in the Sixth Framework Programme's call for an NoE in "Systems Beyond 3G". However, another contender originating from the COST track had valuable expertise in the radio channel and network areas. Bearing this in mind, the NEWCOM++ management made the brilliant move to merge the best of both proposals.

What will be the future of wireless communications? This book indicates possible technology paths to this future. It would be great to give specific answers, but I can formulate only questions. Who will dominate the value chain? Content, carriers, or vendors? I would not go as far as calling the *apps* business a parasitic industry, but, in the words of a non-European service industry leader, "We're continuing to invest to carry the data but we're not really monetising it and this is a global issue that we have to address." The apps hype will lead to more important questions about our way of handling them (or not). Brain research and psychology tell us that humans are not very good at time-sharing, but "always on" forces us into that direction. Will we be able to resist downloading every piece of information offered to us? Will we stay free of gadgets, which are so extremely useful that we yielded to them without realizing that we already have become addicted? Isn't there a hazard to lose skills that in our evolution had been essential for survival? Assuming that many of our daily activities will be handled by apps from out smartphone – or whatever it will be in the future – what will happen, if there appears this tiny little bug disabling the phone? Or it runs out of power at a place where we cannot reload our battery?

In the past, the vast majority of forecasts about numbers of wireless communication users had been consistently on the low side. Recently, I read about a forecast of 50 billion (!) connections by wireless world-wide. Machine-to-machine evidently will have to make the majority of these. Will we communications engineers succeed to reduce the power requirements of these connections to levels that power engineers will be able to supply?

Looking at today's European wireless industry, I see dark clouds hovering. Of course, as wireless communications has become a commodity issues other than scientific ones prevail, by necessity mostly commercial ones. Instead of regarding research as an asset to make future business, inflated controlling and marketing departments at vendors and operators consider research merely as an element of expenditure. Handing out subsidized phones and implementing flat-rate policies ("1000 minutes for free") have given the public the notion that wireless isn't worth much, while shaking the commercial basis of operators. Despite obviously fierce competition, which turned into nothing less than a suicidal fight for market shares, sector-specific regulation still applies for operators – a singular atavism in Europe's neoliberal policies. In contrast, parasitic industries flourish, like selling esoteric plastic chips (without any effect at all) claiming protect against the never-proven hazards of cell-phone radiation.

What can we learn from this brief recount? First, practising cooperation and sharing knowledge is the best way to promote engineering, innovation, and science. The success of NEWCOM++ success is vivid proof. Second, as scientists/engineers are replaced by marketers, it will require new scientific visions to regain Europe's lead in wireless communications. And third, communications engineers, who have a proven record to enhance mankind's abilities, will have to focus on responding to other basic personal needs, as there is security, recognition, possibly play/entertainment, without putting at stake our autonomy.

Enjoy reading about the visions of NEWCOM++!

Vienna, May 2011 *Ernst Bonek*

Acknowledgements

This book is the result of an *integrated* effort of many excellent European researchers all inhabiting for a few years a one and only large virtual laboratory: the European *Network of Excellence in Wireless COMmunications* NEWCOM++ supported by the European Commission's research Framework Program 7. The main objective of a Network of Excellence (NoE) is creating an environment to foster cooperation across the different excellent research Institutions in Europe, and nurturing an attitude towards *integrated research* in a new generation of young researchers. In particular, NEWCOM++ started in January 2008 and ended in April 2011, and its purpose was addressing medium/long term, complex, interdisciplinary, fundamental research problems in the field of wireless networks, with a specific focus towards identifying, casting into the right modeling perspective, and at least partially characterizing the information-communication theoretical limits.

This "Vision Book" is the indirect results of the many researchers who contributed to the success of the NoE, and whose invaluable work we would like to explicitly acknowledge here. The reader will see their names at the start of the papers they contributed to prepare – we just say that it was a pleasure coordinating such a distinguished group of people. But we would also like to explicitly mention the Technical Officers of the European Commission, DG Information Society, who helped us shape, support, and sustain NEWCOM++: Peter Stuckmann and Andy Houghton whose precious and continuing cooperation was the soil on which our research rooted and grew.

In addition to the contributors that appear at the beginning of each paper, a special mention goes to Giacomo Bacci from the University of Pisa whose help in collecting all the contributions, giving them a uniform and coordinated shape and assemble what could be called "a manuscript" was invaluable.

Seeing so many young researchers working together and wishing to pursue a career in research was the best reward we could have asked at the time when we started NEWCOM++. As an encouragement to all those who still wish to pursue a career in research, we conclude this acknowledgement section with a quote by the Danish-born scientist Niels Stensen (1638–1686, Latinized *Nicolaus Steno*), discoverer of the *ductus stenonianus* (the duct of the parotid salivary gland):

Pulchra Sunt Quae Videmus
Quae Scimus Pulchriora
Longe Pulcherrima Quae Ignoramus

or, in plain English,

“Beautiful is what we see
More beautiful is what we know
Most beautiful by far is what we still ignore”

Sergio Benedetto
Luis M. Correia
Marco Luise

Contents

List of Contributors

Carles Antón-Haro
CTTC, Technology Center of Telecommunications of Catalogna, Barcelona, Spain

Erdal Arikan
Bilkent University, Ankara, Turkey

Gerd Asheid
RWTH, Aachen, Germany

Marc Belleville
CEA-LETI, Commisariat à l'Énergie Atomique-Laboratoire d'électronique des technologies de l'information, Grenoble, France

Hanna Bogucka
PUT, Poznan University of Technology, Poland

Giovanni Corazza
CNIT, Consorzio Nazionale Interuniversitario per le Telecomunicazioni, Bologna, Italy

Ioannis Dagres
IASA, Institute of Accelerating Systems and Applications, Athens, Greece

Davide Dardari
CNIT, Consorzio Nazionale Interuniversitario per le Telecomunicazioni, Bologna, Italy

Mischa Dohler
CTTC, Technology Center of Telecommunications of Catalogna, Barcelona, Spain

Pierre Duhamel
Supelec, Higher School of Electricity, France

Yonina Eldar
Technion, Israel Institute of Technology, Tel Aviv, Israel

Laura Galluccio
CNIT, Consorzio Nazionale Interuniversitario per le Telecomunicazioni, Catania, Italy

Sinan Gezici
Bilkent University, Ankara, Turkey

Lorenza Giupponi
CTTC, Technology Center of Telecommunications of Catalogna, Barcelona, Spain

Alessandro Guidotti
CNIT, Consorzio Nazionale Interuniversitario per le Telecomunicazioni, Bologna, Italy

Christine Guillemot
CNRS/INRIA, National Center of Scientific Research, France, National Institute for Research in Computer Science and Control, France

George Iosifidis
IASA, Institute of Accelerating Systems and Applications, Athens, Greece, University of Thessaly, Volos, Greece

Michel Kieffer
CNRS/L2S, National Center of Scientific Research, France

Adrian Klicks
PUT, Poznan University of Technology, Poland

Marios Kountouris
CNRS, National Center of Scientific Research, France

Iordanis Koutsopoulos
IASA, Institute of Accelerating Systems and Applications, Athens, Greece, University of Thessaly, Volos, Greece

Alessandro Leonardi
CNIT, Consorzio Nazionale Interuniversitario per le Telecomunicazioni, Pisa, Italy

Marco Luise
CNIT, Consorzio Nazionale Interuniversitario per le Telecomunicazioni, Pisa, Italy

Enrico Magli
CNIT, Consorzio Nazionale Interuniversitario per le Telecomunicazioni, Torino, Italy

Carles Navarro Manchon
AAU, Aalborg University, Denmark

Flavia Martelli
CNIT, Consorzio Nazionale Interuniversitario per le Telecomunicazioni, Bologna, Italy

Guido Masera
CNIT, Consorzio Nazionale Interuniversitario per le Telecomunicazioni, Torino, Italy

Javier Matamoros
CTTC, Technology Center of Telecommunications of Catalogna, Barcelona, Spain

Natalia Miliou
IASA, Institute of Accelerating Systems and Applications, Athens, Greece

Giacomo Morabito
CNIT, Consorzio Nazionale Interuniversitario per le Telecomunicazioni, Catania, Italy

Dominique Morche
CEA-LETI, Commisariat à l'Énergie Atomique- Laboratoire d'électronique des technologies de l'information, Grenoble, France

Chistophe Moy
CNRS/Supelec, National Center of Scientific Research, France, Higher School of Electricity, France

Dominique Noguet
CEA-LETI, Commisariat à l'Énergie Atomique- Laboratoire d'électronique des technologies de l'information,
Grenoble, France

Gabriella Olmo
CNIT, Consorzio Nazionale Interuniversitario per le Telecomunicazioni,
Torino, Italy

Sergio Palazzo
CNIT, Consorzio Nazionale Interuniversitario per le Telecomunicazioni,
Catania, Italy

Jacques Palicot
CNRS/Supelec, National Center of Scientific Research, France,
Higher School of Electricity,
France

Raffaella Pedone
CNIT, Consorzio Nazionale Interuniversitario per le Telecomunicazioni,
Bologna, Italy

Jordi Pérez-Romero
UPC, Polytechnic University of Catalogna,
Barcelona, Spain

Andreas Polydoros
IASA, Institute of Accelerating Systems and Applications,
Athens, Greece

Venkatesh Ramakrishnan
RWTH, Aachen,
Germany

Shamai Shlomo
Technion, Israel Institute of Technology,
Tel Aviv, Israel

Piotr Tyczka
PUT, Poznan University of Technology,
Poland

Luc Vandendorpe
UCL, Universitè Catholique de Louvain-la-Neuve,
Belgium

Alessandro Vanelli-Coralli
CNIT, Consorzio Nazionale Interuniversitario per le Telecomunicazioni,
Bologna, Italy

Claudio Weidmann
FTW, Telecommunications Research Center,
Vienna, Austria

Andreas Zalonis
IASA, Institute of Accelerating Systems and Applications,
Athens, Greece

Introduction

Vision books are drafted in several contexts (international research fora, such as the Wireless World Research Forum, technological platforms, such as the European e-Mobility, specialised agencies, both profit and no-profit, etc.) and serve different purposes.

By definition, they need to incorporate some "visionary" germs, and as such deal with medium-long term predictions. Some deal with the scientific/technological evolution, trying to predict the future of key enabling technologies by extrapolating today characteristics and up-to-date research trends. The authors are typically renowned scientists and industry leaders, and their reports are mainly aimed at influencing the policy makers at large so as to gain resources in their broad area. Other focus on the societal needs of technology, and try to depict the future needs of prospective customers so as to offer business hints to equipment manufacturers and service providers. A few envision a more comprehensive view of the future shaped by technological advances and its impact on the large scale societal behaviour, offering matter for sociologists' analysis, and, to a more diffused scale, making human beings dream about the impossible to become possible and available.

In their attempt to predict the future trends, the vision books typically undertake one of two approaches: either they present easily foreseeable evolution of present technologies and applications, in which case of course the depicted scenario will become real in a short-medium time frame, or they launch themselves in the realm of disruptive technological breakthroughs and "killer" applications, where almost invariably they fail in their visionary scope. There are many past examples of this incapacity to capture the great/little potential of scientific/technological advances: fibre optic technology (this year Nobel Prize recognised its pioneer, late as usual), the Internet protocol, the Shannon theory, the videotelephony, the cellular SMS, etc.

The reason may be very simple: the great jumps ahead in human history are the result of revolutionary theories/technologies, rather than the evolutionary stretching (and as such, foreseeable) of the existing. They amount to discontinuities, a kind of "nonlinear" progress that can be predicted by the fantasy of novelists like Jules Verne rather than by realistic scientists well rooted in their time.

Benedetto S., Correia L.M., Luise M. (eds.): The Newcom++ Vision Book. Perspectives of Research on Wireless Communications in Europe.
DOI 10.1007/978-88-470-1983-6_1, © Springer-Verlag Italia 2012

For this, and other reasons related to our particular and somewhat narrower perspective, this report is not motivated by the ambition to present a future, comprehensive scenario of the wireless communications in the decades to come. Rather, a community of researchers, shaped under the common ground of a mainly academic network of excellence, have tried to distil their scientific wisdom in a number of areas characterised by the common denominator of wireless communications, by identifying the medium-long term research tendencies/problems, describing the tools to face them and providing a relatively large number of references for the interested reader.

The identified areas and the researchers involved in their redaction reflect the intersection of the major topics in wireless communications with those that are deeply investigated in NEWCOM++; they are preceded in the first paper by an original description of the main trends in user/society needs and the degree of fulfilment that ongoing and future wireless communications standards will more likely help achieving.

The document reflect the ideas and work of many NEWCOM++ researchers (most of them, in fact), and it has been written in layers. For each section an editor has been identified, who has delineated the content for the section and identified the most competent NEWCOM++ researchers to write the particular subsections. At the end, all editors have done a careful work of homogenisation of each section, while the two internal reviewers (the WP leader and the integration activity leader) have supervised the whole work and smoothed the contrasts/repetitions among the various sections.

The following is a concise anticipation of what the NEWCOM++ community has identified as major social applications (and concerns) and technological trends enabling them.

Applications:

- Environmental fitness;
- Social interactions;
- Working environment;
- Entertainment;
- Personalised, location-aware services;
- Health and wellness;
- Security and privacy.

Technological trends:

- Heterogeneous, self-organising, cooperative and cognitive networks;
- Global system optimisation adopting a multilayer approach;
- Adaptive, shared use of the bandwidth through cognitive radio;
- Enhancement of bandwidth efficiency (up to the information-theoretical limits) through:
  - Source, channel and network coding;
  - Adaptive techniques in radio resource management;
  - Concentrated and distributed MIMO;

- 3-D multimedia: coding and faithful delivery;
- Flexible, energy-aware hardware technologies;
- Global attention to energy saving: "green ICT and ICT for green".

The first paper offers a view on trends currently characterizing the ICT world, with special focus on wireless technologies. It first attempts at classifying personal needs that drive the growth and evolution of ICT trends and technologies, and then it presents an overview of ICT trends and a concise description of cutting-edge standards and technologies in the ICT wireless context, providing for each a mapping onto ICT trends, and showing which trends are embodied by the technology and which personal needs can be satisfied by it.

The second and third paper are devoted to the future of wireless networks, with emphasis on features that are likely to characterise their evolution: heterogeneity and opportunism (the second one), and cognition and cooperation (the third one). In particular, the second paper starts with the implications of network heterogeneity on how wireless networks will be operated and perceived by the users, with the corresponding requirements for smart radio resource management strategies. Then, it addresses the concept of self-organising networks, and it describes opportunistic networks, vehicular communication networks and the Future Internet and the Internet of Things concept. The third paper approaches the concept of cognitive radio networks, in its aspects of spectrum sensing, spectrum management, spectrum sharing and cognitive positioning, a feature that opens the door to the series of location-aware wireless services. The second part of the paper deals with cooperation in wireless networks, a concept that extends the capabilities of multiple-input multiple-output (MIMO) techniques to various forms of non-colocated, yet cooperating, antennas, such as those of different users and/or base stations. Finally, it introduces the innovative concept of "docitive" radio and networks, characterised by a certain degree of intelligence which allows them to work properly under conditions they were initially not designed for.

The fourth paper stems from one of the key work packages of NEWCOM++, investigating the information-theoretical ultimate limits of wireless networks through mathematical tools like game theory, stochastic geometry, and random graphs, and techniques to approach them such as decentralised network control, coordination and competition strategies, network coding. The section concludes with a future perspective on wireless sensor networks.

In the fifth paper the emphasis is on the maximization of the throughput per unit of bandwidth through adaptive strategies in the radio access. The first part of the section discusses the theoretical background of the problem and proceeds with a discussion of coding/decoding techniques, as well as techniques that employ adaptive modulation and coding. A more detailed discussion of coding/decoding techniques and, in particular, iterative techniques for wireless receivers is the topic of the second part of the section. The last part discusses cognitive spectrum usage with a special focus on applied game theory for this goal.

The sixth paper refers to processing and reliably delivering of multimedia signals. As the most challenging future trend, the main features of capturing, processing and delivering 3-dimensional signals are described in their present status and future evolution. Network coding is identified as one of the key enabling technologies to the global optimisation of the wireless network and to increase the throughput, reduce the delay/latency, improve the quality and reduce the energy consumption. The final part of the section is devoted to two deeply studied but yet to fully exploit their potential, like cross-layer optimisation and joint source and channel coding.

The seventh paper is fully devoted to a deep insight into the present and future of "hard" technologies for wireless communications. It starts with the trends in silicon technologies, and continues with the analysis of conflicting requirements such as high computing power, great flexibility, and power efficient design. The approach of software defined radio and its challenges/opportunity is then described, in its present status and future evolution, with a brief mention of the radio-frequency section of the receiver.

Finally, the eighth paper deals with "green" communications, a concept drawing from the universally shared concern about climate changes and aspiration toward a sustainable development guided by ecological considerations. ICT at large, and communications in particular, have been recognised as a very significant contributor to the energy consumption and, thus, CO_2 emission. The section presents a holistic approach to the sustainable development of wireless communications, based on the idea that all wireless system and network as well as all the infrastructure components must contribute to the overall sustainability. Then, it deals in particular with the main tools of green communication, such as flexible and cognitive radio for energy efficiency and a better spectrum usage and reduced pollution.

Sergio Benedetto
Luis M. Correia
Marco Luise

Trends in user/society needs and future wireless standards

Giovanni Corazza (✉), Alessandro Vanelli-Coralli, Raffaella Pedone, Andreas Polydoros, Dominique Noguet, Adrian Klicks, Jordi Pérez-Romero, Alessandro Guidotti, and Flavia Martelli

This section offers a view on trends currently characterizing the ICT world, with special focus on wireless technologies in line with the NEWCOM++ research objectives. To approach this complex task, the activities carried out within WPI.6 are considered as a reference. At first, an attempt to classify personal needs is provided in order to present the rationale towards which ICT trends and technologies are growing and evolving. Then, an overview of ICT trends is presented through an original metaphor of the current ICT picture and biological DNA. Continuing our analysis, cutting-edge standards and technologies in the ICT wireless context are recalled, providing for each a mapping onto ICT trends, showing which trends are embodied by the technology and which personal needs can be satisfied in this manner. Finally, we conclude this section discussing the impact of ICT pervasiveness on society, human relations, and even individual psychology: a new phenomenon is created, which we identify as the *information divide*, which includes both the *digital divide* affecting all those who have no or limited access to the Internet, as well as the *psychological divide*, related with the information overload which impacts upon all those who do have access [1].

1 Personal needs

Describing the picture of trends and user/society needs in the ICT European context is not a simple task. The first step we take in this attempt is to consider the society conditions in which the ICT evolution and innovation are currently taking place: in this way, the focus is firstly put on the individual, in line with the approaches undertaken in parallel by other policy programmes in Europe, among which it is worth citing the AAL (Ambient Assisted Living) programme as the champion for focused attention on the human being 0.

The individual person is always at the centre of any societal development because any transformation translates necessarily into modifications of personal habits, and often is actually resting upon their intrinsic nature. Re-interpreting in a more modern

Benedetto S., Correia L.M., Luise M. (eds.): The Newcom++ Vision Book. Perspectives of Research on Wireless Communications in Europe.
DOI 10.1007/978-88-470-1983-6_2,

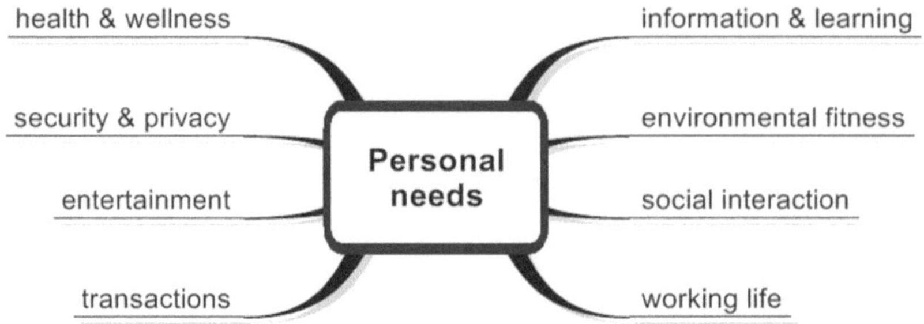

Fig. 1. The map of personal needs

key the classical approach by Maslow [2] who modelled the personal necessities through a universally famous pyramid, needs can be grouped into 8 major categories as described in the map in Fig. 1. Of course, classification of human needs under these discrete classes is not a simple task, with the unavoidable result that often an overlap between different needs can be detected, a fact that is inherently related to the complex nature of human beings coupled with the degree of subjective interpretation that is different for different people.

It is interesting to note that each of these 8 categories of needs can be linked to a series of clear trends in the current ICT European arena. Accordingly, insight into this classification is a useful starting point to guide us towards the identification of the ICT trend roadmap.

1.1 Information and learning

This is the natural need for knowing, being up-to-date and informed anytime and anywhere, which has assumed a particularly compelling connotation in recent years with the rapid and capillary diffusion of Internet applications. In the digital era, the person is more and more eager to reach rapidly and efficiently all contents he/she needs in a precise moment, at home, at work, on the move. This trend is confirmed by the interest towards the creation of *digital libraries* in the current European research framework programme, and by the advent of new applications aiming at satisfying the user thirst for receiving information and learning in remote areas, in the form of e-learning. Notably, these trends are first responses to the social challenge of inclusion, which contributes in avoiding person isolation from society through e inclusion mechanisms. As broadband becomes a necessity of daily life (*broadband for all*), the impact of information exclusion for citizens that do not have broadband access or who cannot afford it will be dramatic. Today's digital divide may become tomorrow's social-exclusion.

1.2 Environmental fitness

This is a broad interpretation of the necessity for any human being to adapt the surrounding ambient to its habits in order to allow comfortable living. This need is mapped onto a series of ICT trends that can be identified in the current picture. First of all, *mobility* reflects the possibility for the user to communicate, work, be entertained, and more generically reach the outside world while being on-the-move, a very popular and wide-spread tendency among users of all kinds (from consumer to professional and institutional). Ubiquitous multimedia communications have been the key to third generation cellular standards, and foreseen to become even more attractive in the next future. The current trend towards *info-mobility* is a specification of the wider mobility concept, consisting in the distribution of information to passengers of various transport means. The most relevant example is that of car passengers, which opens the way to more sophisticated applications such as Automatic Vehicle Monitoring or *intelligent cars* (the Intelligent Cars Initiative, i2010 [3]). *Ambient awareness* is another form of environmental fitness, integrating aspects of ambient intelligence into context awareness systems. A key role is played by navigation systems, such as GPS or the European Galileo, which enable location-based services. In this scenario, ambient awareness consists in computer mediated communication systems that help people maintain a peripheral awareness of others, translating into connections between households and mobile individuals, with automatic capture capabilities of awareness information and with a pervasive topology in the environment. Finally, the process of adaptation of the surrounding environment to the user ICT experience should avoid to forget the environmental footprint, identifying solutions to exploit resources efficiently and favour sustainability, in line with the *Green IT* and *energy efficiency* requirements.

1.3 Social interaction

This is the natural need of humans to belong to some form of society, from a small scale (a family, a group) to a larger scale (a community, a Country), in response to the continuous search for support, discussion, comparison and opinion exchange with other persons. This need takes a completely new connotation with the recent technology advances. In particular, it is worth mentioning here the tendency for *social networking*, so highly popular nowadays amongst younger generations, and the trend towards virtual interactions in *cyber-realities* where all aspects of life can be artificially re-constructed, through the creation of virtual identities or avatars (Second Life). The web is now used in a participative mode, as testified by the success of a number of applications such as Wikipedia, blogs, MySpace, Facebook, YouTube, GoogleMaps, etcetera. In this framework, other foreseen trends are *enhanced reality* (with brain electrochemical stimulation), and *augmented reality* (combination of real and digitally modified identities and environment). Of course, although we focus here on the more technologically oriented evolutions, this does not mean that real interactions among human beings should ever be replaced by their artificial counterparts.

1.4 Working life

This is the need for self-fulfilment and self achievement realized through working activities, which enables to exploit talents and education, contributing fundamentally to self-esteem. The need for a satisfactory working life is reflected in an important ICT trend, that we can identify as *e-Business*, comprising home-based or mobile-based teleworking, broadband connection between different enterprise premises, remote training, videoconferencing, etcetera. The nomadic use of ICT will challenge the meaning of 'being at work'.

1.5 Transactions

This is the need of humans to be supplied with services and goods in order to satisfy material/immaterial desires. ICT can be of help here with the supply of a host of *e-commerce* tools, that can be used in business, private, or consumer contexts. Home banking is but one example.

1.6 Entertainment

This is the need for amusement and hobbies, to aliment the innate human tendencies towards recreation. In partial overlap with the social interaction need when applied to the personal sphere, the necessity for entertainment is one of the most typical goals of ICT services, taking on various forms such as on-line gaming, portable consoles, mobile TV, P2P downloads, MP3 players, social networking (YouTube, Facebook, SecondLife), etcetera. Indeed, social networking is a direct response to the entertainment need.

1.7 Security and privacy

These are primary needs of any person to feel protected during any aspect of life. In particular, this translates into the need for preserved secrecy on private data, which becomes a real social challenge in an era where communication of personal digital data is at the centre of many applications, especially web-based Internet applications. The success of social networking websites is the most evident example of the dissemination of private data over the Internet; search engines get and store data after every query, the content of personal e-mails is stored in remote servers and in some cases scanned for commercial and security reasons. Moreover, the user-friendly tools typical of the new, second-generation Web have made the Internet a platform for a growing number of financial transactions, ranging from purchases of goods to money transfers between electronic bank accounts. In the digital economy, public authorities have the responsibility to make sure that citizens can trust the use of the Internet, combating cybercrime, avoiding episodes of personal identity theft and of malware (malicious software) dissemination via spam and website attacks, and improving at the same time the authentication, validation and *digital identity management* for high level transactions (Internet banking).

1.8 Health and wellness

This is the last but fundamental need of humans for truly satisfactory livelihood, which should be provided anywhere and anytime thanks to the ICT enablers. As a prominent example, *e-health* applications remotely assist patients, for example exploiting body area networks capable of monitoring vital parameters [4]. E-heath ICT is the response to critical social challenges such as *home care* (or domiciliary care), i.e. health care or supportive care provided in the patient home by healthcare professionals, and *independent living*, i.e. supporting assistance to disabled people. All of this at a significantly reduced social cost.

ICT can indubitably be considered as an enabler for satisfying these user needs via different forms of response, as outlined above. Europe is investing considerably in research activities in these fields, to devise novel and efficient applications in response to the user primary needs and to solve the social challenges posed by continuous evolution and innovation.

2 Major trends in the ICT world

Taking on an original point of view, we here present what we believe are the underlying and unifying major trends in ICT arena. We have seen that there exist many trends, in response to various personal needs, and it is difficult or even unnecessary to give a ranking, but it is possible and useful to try to classify them in order to make the discussion more interesting and organized [5].

We elect to say that the two main forces which are governing the revolution brought in by ICT technologies are *personalisation* and *distribution*. Using a biological metaphor, we can think of personalisation and distribution as the two strands, the two filaments, in a DNA structure, upon which and through which other trends are formed. This is what we call *the DNA of ICT evolution*, represented pictorially in Fig. 2. The two filaments generate and are linked by the bases, each one corresponding to a major ICT trend. Therefore, in the following we will identify *personalisation* and *distribution* as meta-trends, or “trends of trends”. Before we get into the more specific elements of the DNA of ICT evolution, let us dwell on personalisation and distribution to give a rationale to our statement on their importance.

2.1 The personalisation meta-trend

The success of the industrial society was based on the mass production of goods at low cost, to be sold to a consumer market with homogenized tastes and desires. This paradigm today is completely reversed. The strategy is to go to the person, produce for the individual, satisfy specific needs, segment the market into small niches each tailored to a particular group of persons. This is all made possible by the fact that today it is feasible to produce at low cost with flexibility and modularity. Now, this trend is clearly reflected in the world of communications. Since the advent of mobile telephony, also identified as *personal* communications, we do not call a

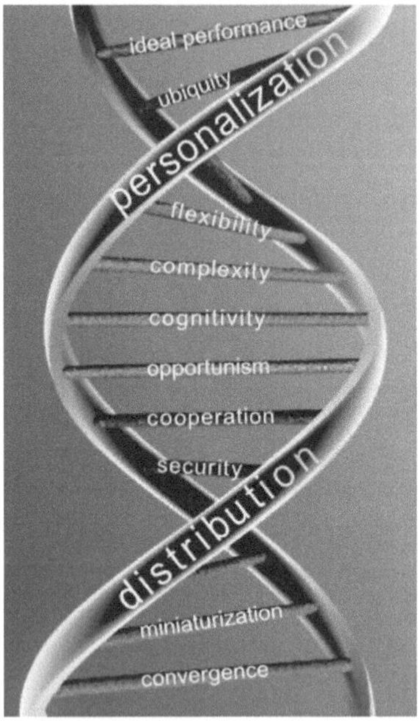

Fig. 2. The DNA metaphor of the ICT trends

location, but an individual. We are now used to see the number that is calling us, associated to a person in our phonebook, and we decide whether to take the call or not. We select a specific tariff structure, suited to our calling profile, and we may even have lower tariffs when we call specific groups of people. In terms of accessing the Internet, we start from our preferred home page, we can browse over portals shaped according to our own profile, we can select our favoured content from a huge offering, we can even become content producers by uploading pictures and movie clips. The individual has become a source of information, and not only a sink to be filled with advertisement. Even when watching TV, we now have at our disposal a growing number of on-demand offerings, from which we can freely select. The individual is more important, enjoys more freedom, is much more active than it was in the past.

2.2 The distribution meta-trend

Another fundamental ingredient of the industrial society was the concentration of resources and intelligence into a few centres. This made interactions and investments more efficient, while also increasing the risk associated to losing a centre, and caused large movements of people (the first of which was the move out of the country into cities). The structure of companies was based on rather rigid pyramids, with very specific work functions and descriptions, and linear work flow procedures. Again, also this paradigm is reversed in the information society. Today, there is a clear trend towards the distribution of resources. In the world of manufacturing, it is customary to have parts produced in geographically distant premises. This reduces costs and creates a global economy where effects propagate unboundedly. Intelligence is distributed and decisions are obtained through a network of interactions. This adds greatly to the responsibility of each person in the organization, and the work functions become ever more flexible. This also creates a need for continual education, for it is not possible to adapt to the fast dynamics of the current societal evolution if one considers that his/her training ended in school. Companies' structures evolve from pyramids into networks of intelligent nodes, and the structure may evolve on a per product basis. The more futuristic version being the virtual company, where different entities join into specific ventures which only have the lifetime of a product life cycle. It can be stated, without the minimal shade of doubt, that all of this is resting upon ICT technologies. The network of people relies on the telecommunication network, and in the future, the same will happen for the network of *things*, also known as machine-to-machine communications. A distributed society, a distributed company, need to find the necessary information anywhere they might reside. This is only possible thanks to the Internet and the associated search engines which allow to have nearly all the information in the world at one's fingertips. Intelligence is pushed to the edges, which reduces the risk associated to the loss of any network node, but at the same time requires a new ethical code, as well as security in communications. Networks evolve from heavy infrastructures to lightweight ad-hoc self-organizing topologies, where the role of operators needs to be defined anew.

As in the DNA structure, the two fundamental strands of *personalisation* and *distribution* are actually running into opposite directions, i.e. they are the anti-parallel support of the ICT DNA double helix. In fact, while personalisation implies a local view, distribution naturally translates into a global reality. Therefore, ICT is imposing on society a complex organization, whereby the global truth is formed through belief propagation from individual nodes which intelligently work within their local boundaries. As we will see later, this has much to do with the concepts underpinning iterative processing algorithms.

2.3 ICT trends

The two meta-trends of personalisation and distribution can be combined in various ways to form the trends of ICT evolution, which in our DNA metaphor correspond to the bases, the sequence of which identifies the genetic code. The following is our elected list of ten trends: Ideal performance, Ubiquity, Flexibility, Complexity, Cognitivity, Opportunism, Cooperation, Security, Miniaturization, Convergence. Let us dwell on each in turn.

2.3.1 Ideal performance

The search for the ultimate ideal performance is the major force behind the evolution of any technical or technological system. It appears to be in the nature of the human kind to strive for the extraction of the maximum possible output, the optimal exploitation of resources, with the largest possible efficiency. This requires knowledge of the ultimate performance boundaries. In the case of communications, the boundaries are the object of Information and Communication theories, which in many instances do indicate where these limits are. The trend is therefore towards achieving the performance limits set by Information Theory. How does this link to the two meta-trends? The optimization of performance requires perfect fit to the specific communication conditions (propagation channel, interference, transmission format, etcetera), which can be interpreted as matching the individual user needs and constraints. This is part of *personalisation*. On the other hand, the limits set by Information Theory are not restricted to a single link, but can and should be extended to the more complex case of networks. In this case, achieving the optimal limits requires global optimization, to balance fairness and overall throughput, with distributed intelligence. This is evidently part of the distribution meta-trend.

2.3.2 Ubiquity

The trend towards ubiquitous communications, the overused "Anywhere, anytime" motto, has been the driver for the evolution of cellular communications since the 70's. Coverage is today extremely good in most urban areas, and surprisingly good in unexpected locations, even though obviously gaps remain in developing parts of the world. What is yet necessary is to sharply increase the geographic spectral efficiency (in bit/s/Hz/km^2), to provide ubiquitous *broadband* wireless access. Associated to ubiquity, we find mobility and pervasiveness. Mobility is the trend towards communication systems which can interconnect terminals moving at any speed, including all types of vehicles, trains, airplanes, ships. Pervasiveness is the trends towards finding connectivity all around us, in a truly *wireless ambient*. This concept has many important social implications, as already discussed above. Essentially, by living into a collaborative wireless ambient, the individual can benefit from optimized environmental fitness, which satisfies a basic human need. This is clearly a specification of the *personalisation* meta-trend. At the same time, this personal fitness can be carried along in any location, becoming ubiquitous fitness, an evident derivation from the *distribution* meta-trend. Even though ubiquity and distribution may seem very similar concepts, we separate them by limiting the interpretation of ubiquity to the pervasiveness of wireless networks, and by attributing to distribution this and all other implications related to social aspects, work organization, global economy, etcetera.

2.3.3 Flexibility

Along with the search for Ideal performance, this is also a major trend in the evolution of any technical system. All engineering systems start as rather simple and rigid, performing but a few functions, with limited scope for modifications in response to user needs. In the course of its development, the system acquires more and more functions, more and more options, which can be selected flexibly depending on instantaneous necessities. This is a very strong trend in wireless communications. Transmission systems, protocols, and terminals, are being designed as reconfigurable entities, with capabilities that can be flexibly adapted to the conditions set by the propagation channels, the transmission buffers, the spectrum availability, the interference environment, the desired quality of service, etcetera. Dynamic spectrum assignment strategies are being devised and starting to find their way into regulatory policies. Digital electronics capabilities are exploited to design software radios and flexible radios. Even analogue electronics is now being bent to the requirements of designing flexible RF front-ends, with reconfigurable filters over large bandwidths. It is an easy task to map the trend towards *flexibility* as a direct son of the *personalisation* meta-trend. In fact, it is obvious that flexibility is only useful if it is used to accommodate individual conditions and needs. On the other hand, it may be harder to describe the connection with the distribution meta-trend. However, *flexibility* at system level requires knowledge of all local conditions, in order to find a global optimum satisfying the requirements of the entire user population. Therefore, we can say that *global flexibility* is related to the trends towards *distribution* of intelligence, where as a minimum each user must sense its own environment and feed back this information to peers or to base stations. Also, the trend towards flexible network topologies is clearly out-spinning from the *distribution* meta-trend.

2.3.4 Complexity

Technical systems always evolve towards increasing levels of complexity, as functionalities increase and performance improves. On the other hand, technological complexity can become a major hurdle in its usability. Therefore, while internal complexity increases monotonously, there is a contextual trend towards the simplification of the human-to-machine interface. Complexity and simplicity must live together in harmony. As complexity grows, we must face the danger of increasing energy consumption, which could make entire systems unsustainable. A clear trend towards the design of "green" technical systems is growing powerfully nowadays. We could say that energy consumption is to complexity as energy saving is to simplicity. Mapped onto the world of wireless communications, complexity is visible in systems, protocols, terminals, network equipment, essentially in every element. The need to simplify is stringent for user terminals, but also for network management. And we can say that "green" communications are emerging as very hot area of research and development. The relationship between complexity and the *personalisation* meta-trend is inherent in the fact that we do not accept standard and rigid solutions, but rather we always look for configurations which are adapted to individual needs. A personalised solution is always more complex than a standard item. The key enabler for the realization of complex systems is the fact that today we are able to produce personalised objects in a very cost effective manner. It is also true that, in many instances, personalisation is perceived by the final user, but it is in reality a specific combination of a few standard objects. In view of the *distribution* meta-trend, it should be apparent that distributing intelligence, responsibilities, management functions, all translate into a more complex system. In this case, complexity also brings in the concept of *emergence*: the arising of novel and coherent structures, patterns and properties during the process of self-organization in complex distributed systems. Emergence can be weak when it can be reduced to its elemental parts, or strong when irreducible. Irreducible emergence can be thought of as an independent system, living a life of its own.

2.3.5 Cognitivity, self-organization and bio-inspiration

As complexity of systems grows larger, control becomes more and more difficult. At a first inspection, it would be desirable to be able to set rigid rules to which all system elements should abide. This has worked in the past and still does today. However, this can only be pushed to a limit, when exceptions to the rules become frequently necessary, but difficult to handle, and the overall efficiency is severely degraded. Also the system may become extremely large, and scalability of control becomes a major issue. Or, finally, flexibility demands may pose tremendous challenges to setting correct and efficient rules. In front of all of these difficulties, we are turning our observations to nature, where incredibly complex beings live apparently without any form of rigid control. This is the source for bio-inspired algorithms, techniques, and protocols. We see that life in nature is self-organized, and we can try to apply self-organization into devices, networks, and systems. And clearly, the most beautiful and powerful example of self-organized system is the human brain, with its capability of cognition. Therefore, the extreme finalization of this trend is to endow devices, networks and systems with cognitivity, i.e. the cognition capability. Hence the example of cognitive radio, where radio spectrum is not assigned a priori, but is cognitively selected based upon observations of the wireless environment. Seen through the light of the *personalisation* meta-trend, we can see that we are actually turning our network nodes and devices into primitive forms of "persons", with a certain amount of artificial intelligence, which allows them to "think" and make decisions with a certain degree of autonomy. In the Internet of Things, the human element largely disappears, and the network is completely populated by artificial beings, or agents, which carry out functions to achieve specific objectives. The *distribution* meta-trend is related very closely to the concept of self-organization, where local realities and decisions contribute, through message passing, to the global behaviour. This can be brought to the extreme where the overall objective functions, such as for example the estimation of a parameter, are elaborated only in a distributed manner, and the final result is not necessarily collected at a fusion centre, but can itself be distributed into the network.

2.3.6 Opportunism

With increasing degrees of distributed intelligence, flexibility, and complexity, it becomes crucial to execute operations not at any generic time instant, but when and only when the conditions are optimal. In other words, it becomes necessary to catch the opportunity for performing a specified task in the most efficient manner, and with the largest associated benefit. Indeed, all systems are dynamically varying in several dimensions (as a minimum, time), which means that conditions will fluctuate and opportunities will be created. To use a dimension or another, depending on the underlying conditions, can be interpreted as a form of diversity. Therefore, opportunism is a way to exploit diversity, choosing from time to time the path which offers the minimum resistance to our action, and thus optimizing the use of resources. In a sense, opportunism can be seen as the opposite of the brute-force approach, where exploitation of resources is total and completely independent of the ensuing conditions. The beauty of this is the fact that, in a network (be it technical or social) the total amount of resources is limited, so if each one use the minimum necessary to achieve its own purposes, then the overall efficiency is maximized. In other words, the use of brute force from any single individual hurts the entire network. In wireless systems and networks, and particularly in the family of Beyond 3G cellular networks, opportunism has become a major flagship for resource assignment, scheduling, multiple access. Resources are given dynamically to those terminals which are at a particular instant enjoying the best channel conditions, which will allow to serve them with the minimum effort and maximal efficiency. In order to avoid that some terminals are always left out of the game, opportunism should always go along with fairness, implemented in one of its several possible embodiments. In this specific case, the *personalisation* meta-trend materializes in the fact that we go after the opportunity which is occurring for a specific individual, knowing that it will only last for a limited window of time. On the other hand, we want to be fair to all users, and as such protocols and strategies are ready to consider also the needs of those for which opportunities do not seem to happen, at least not with sufficient frequency. In terms of the *distribution* meta-trend, we observe that opportunities may also be visible at a local level. This is because, to enable scalability, it is not conceivable that all information be collected in a single decision making node. Therefore, decisions to seize specific opportunities should be taken locally, with feedback on instantaneous conditions transmitted only when and where necessary, possibly on short legs to minimize latency and thus maximize network reactivity. Hence, opportunism must go along with distribution.

2.3.7 Cooperation

Cooperation can be seen as the virtuous consequence of awareness. If an individual, or an entity, is isolated, it can only work for its own specific goals. On the other hand, even if the entity is not isolated, but is unaware of the needs, or even the sole presence, of other entities around it, it will behave exactly as it did in isolation, working towards the achievement of its objectives. Only when an entity becomes aware of the presence, requirements, and needs of other entities around it, then it can realize that working in isolation may not be the most efficient way. Even the objectives are modified, at least because one sees not only its own objectives but also those of others, thus creating the notion of global objectives. Awareness generates a change of perspective, which can lead to various forms of cooperation amongst the individuals. Cooperation requires trust, fairness, and regulation, in order to ensure that all individuals benefit from the process. Cooperation in wireless communications can, for example, take the form of relaying the information sent by another user, in order to help it reach the final destination. In this way, the cooperating node is spending part of its resources not to achieve its own objectives, but rather to help another node do so. In return, it will trust that the situation will reverse when its own opportunity comes along. It is clear then that cooperation and opportunism go together, as the mechanism for cooperation will adapt itself to the underlying conditions which will vary dynamically over time. Other interesting forms of cooperation can be envisaged for virtual beamforming, virtual MIMO, collaborative positioning, etcetera. Cooperation is an act between individuals, and as such it possesses intrinsically the character of the *personalisation* meta-trend. The personalised network entity is aware of the other entities, cognitively decides that it is useful to cooperate, trusts the other entities, expects to receive mutual benefits and to achieve its own goals while contributing to the global goals. This comes very close to the description of the behaviour of a person in a social network. On the other hand, the exploitation of cooperation means, once more, that the operations in the network do not belong to a single terminal and a single central control entity, but rather require the involvement of a multitude of actors, distributed of the area of service, whereby decisions and operation occur as the result of the overall interaction. This is clearly in line with the *distribution* meta-trend, and we can say that cooperation without distribution is impossible, and distribution without cooperation is less effective and not exploited to its fullest.

2.3.8 Security

We must also recognize that, in front of all the positive aspects brought in by the personalisation and distribution meta-trends, there is also an associated increase in the risk of misuse of ICT technologies. Centralized control may be bulky and in some cases unfair, but it can also serve as a guarantee for secure transactions, which can be protected more easily by various kinds of threats. On the other hand, when organizations become distributed, when decision making is the result of consensus, when resource management requires information from the edges, then it is clear that there are so many more possibilities for an alien to come in and disturb or deviate the process far from its intended objectives. And since there is a trend for personalisation, any individual or any entity is up front with all of its features, which can be stolen or misused in many ways. Therefore, the meta-trends of *personalisation* and *distribution* require that much attention is paid to ensure security, guarantee privacy, defy malicious attacks, propagate trust. This applies to society in general, and certainly it does also to wireless communications, which traditionally have been the weak side of network security. One special word for trust: it is not just a matter of making sure that content is encrypted, that access is conditional to authentication, that sensitive data is not exchanged (or at least not frequently). It also a matter to make sure that the final user *perceives* that using ICT technologies is secure. In other words, there must be trust in ICT technologies, or else the uptake will always be below expectations, and the impact much more limited than the potential.

2.3.9 Miniaturization

In the evolution of technology, we always see a trend towards miniaturization, as the results of improvements in the processes and in the understanding of the underlying physics. This has worked perfectly in the case of digital electronics, where the scale of integration of ICs (Integrated Circuits) has grown exponentially through the years. Digital ICs are horizontal enablers for the progress in wireless communications, not only from the technical point of view, but also from the economic side, given that the cost of ICs has also decreased steadily through the years. And the end is not in sight: with the rush for nanoscale devices, unprecedented improvements are yet on their way. Also, circuits built on organic materials promise to change forever the notion of "hardware", as we will be seeing and use devices in plastic and even softer materials. Nanotechnologies will produce entire nanosystems, which can be distributed as smart dust for a multitude of distributed sensing applications. Here comes the relationship with the distribution meta-trend. On the other hand, the trend towards miniaturization can also be seen in a more general way. In terms of cellular mobile networks, there is a clear trend towards the miniaturization of cells, which went from macro to micro, pico, and now femto-cells. The femto-cell is intended to be installed by and individual in a home or a small office, and as such we can see the connection with the *personalisation* meta-trend. We can also see the miniaturization of networks, as for the case of the body area network, interconnecting different parts of the body for various applications for the benefit of the individual, such as e-health.

2.3.10 Convergence

Last but not least, convergence. It is placed at the end because in a way it connects all previous trends. In a general sense, convergence can be seen as that process according to which concepts which used to be separated come together to form new meaning. The process eliminates barriers and distinctions, and creates larger classes. Since we use classification as an instrument for clarification, it is indeed true that convergence always causes a certain amount of confusion, as previous certainties are questioned and new approaches must come in. In economic terms, convergence is an earthquake that shakes market shares and inevitably increases both opportunities and threats. From the point of view of scientific research, convergence can be seen as the uprising of interdisciplinary research, where competences from different areas are merged. The major benefit is that different frames of mind coming together have the potential to produce breakthrough innovation. From the point of view of wireless communications, convergence plays a major role in at least two ways. First of all, the distinction between mobile and fixed telephony is vanishing, with operators offering bundles which include also Internet access and TV (the so-called quadruple play). Considering also the fact that IP (Internet Protocol) is rising as the common network protocol for all services, we can see clearly the incredible force of service convergence. Secondly, there is also convergence in the world of wireless terminals, which more and more become phones, computers, cameras, organizers, etcetera. Let us interpret the trend for convergence in view of the *personalisation* meta-trend. No matter where the person is, we want to provide the same access conditions, the same service profile, as if the person was virtually at home. Convergence of networks and terminals can enable this concept. On the other hand, this can also be seen as the famous "anywhere, anytime" motto, i.e. the fact that we are always surrounded by a converged (albeit heterogeneous) network infrastructure, of which we don't want to know the details, as long as we can use it for our purposes. Therefore, convergence can be seen as an enabler for the *distribution* meta-trend, because without it we would not see a seamless wireless ambient but rather a jigsaw puzzle of technologies.

3 Technologies and standards for next generation ICT evolution

During its second year of activities, WPI.6 has focused on actualizing the scenario described above and identifying the most relevant standards and technologies that are likely to play a central role in the next generation of the ICT evolution. To this aim, these principal domains have emerged to be the cutting-edge:

- cellular systems, where ITU IMT-Advanced sets the main objectives, embodied according to different approaches by the proposals of 3GPP LTE-A (Long Term Evolution – Advanced) and WiMAX;
- mobile ad-hoc networks and body area networks, where standardization activities are on-going within IEEE 802.15 and Task Group 6;
- broadcasting systems, where the focus is specifically put on the most recent developments of the Digital Video Broadcasting standard families, i.e. on DVB-T2 and DVB-SH and of digital radio;
- cognitive radio, where inputs are provided by the activities of the IEEE 1900 standard and the recent vision of OFCOM and ETSI RRS;
- Internet, where novel concepts are being developed towards the construction of the Future Internet; a complete and interesting picture of this evolving trend is provided by the recent activities carried out within the European Commission at inter- ETPs (European Technology Platforms) level.

For a detailed view on these domains and the related technologies, the reader can refer to [6]. In the following, the main wireless technologies examined in [6] are briefly recalled, i.e. LTE-A, WiMAX, MANETs and BANs, DVB-T2, DVB-SH, digital radio, CR, and Future Internet, providing an insight on how each of them manages to answer to the individual needs of Section 1 and embodies the ICT trends of Section 2.

Along with this analysis, an estimate of the time frame characterizing each of these standards is provided. In this attempt, the natural life-time of each technology trend is modelled through three main time phases:

- the first is the ramp up period where technology starts spreading with a diffusion rate that usually begins timid and then increases towards higher penetrations;
- the second is the maturity period, where the diffusion of the technology is almost set to a constant level;
- the third is fade out where the technology penetration gradually decreases, substituted by new competitors.

Besides the time frame, efforts are spent also to forecast the market impact of each technology and to provide a measure of the potential for scientific research underneath each standard, with the objective to guide NEWCOM++ researchers towards the most cutting-edge thematic.

3.1 3GPP LTE-Advanced

3.1.1 Brief description of LTE-Advanced

The LTE-A (Advanced) standard [7] is the new generation of the LTE (Long Term Evolution) standard aiming at designing the evolution of E-UTRA (Enhanced UTRA) and E-UTRAN (Enhanced UTRAN) to meet/exceed the IMT-Advanced capabilities recently set by ITU-R and to satisfy more stringent requirements set by 3GPP operators, encompassing physical layer, Radio interface layer 2 and RRC, E-UTRAN architecture, RF, including aspects of wider channel bandwidth than 20 MHz, and advanced system concepts.

Briefly, LTE-A targets a *downlink peak data rate* of 1 Gbit/s and an *uplink peak data rate* of 500 Mbit/s, with *downlink peak spectrum efficiency* of 30 bit/s/Hz and *uplink peak spectrum efficiency* of 15 bit/s/Hz. The system supports *mobility* across the cellular network for various mobile speeds up to 350 km/h (or perhaps even up to 500 km/h depending on the frequency band). *Spectrum flexibility* is achieved exploiting *wider* bandwidth than 20 MHz via carrier aggregation. MIMO and beamforming techniques are foreseen as advanced techniques to improve performance and resilience against interference.

3.1.2 Individual need satisfaction by LTE-Advanced

In Table 1 we discuss how the LTE-A standard responds to the individual needs identified in Section 1. Needless to say, the interpretation is subjective, but nevertheless helpful in drawing conclusions and understanding what opportunities will come next.

3.1.3 Mapping LTE-Advanced onto ICT trends

LTE-A is strongly related to the ICT trends described in Section 2 as detailed in Table 2.

3.1.4 Time frame for LTE-A diffusion

Even if the proposal for LTE-A has been submitted to ITU in October 2009, it is likely that 3G technology will be the main character of the cellular scene in 2010 and 2011; forecasts can position the beginning of LTE-A diffusion in the telecom market nearly in 2012, year in which the ramp up of LTE-A technology is feasible to take place. This is perfectly in line also with the time schedule foreseen by IMT-Advanced development set by ITU. This first phase of rapid growth of the system can be stated will reach its maturity, assessing an almost constant penetration, in 2015–2016, followed then by a smooth fade out period, when it is likely that a fifth generation cellular system will start to be designed.

3.1.5 Market impact of LTE-A technology

The market impact of LTE-A will likely be extremely interesting; forecasts count up to billions of terminals and subscribers reached during the maturity phase.

Table 1. Individual needs satisfaction by LTE-A technology

Individual need	*Degree of satisfaction of the individual needs by LTE-A technology*
Information and learning	LTE-A development intrinsically respond to the requisite for accessing information anywhere and anytime, enabling fruition while on the move and increasing the supported data rate towards the paradigm of broadband for all
Environmental fitness	Supporting mobility up to very high speed, LTE-A responds to the need for environmental fitness, allowing the user to receive information and access applications from any geographic point
Social interaction	Satisfaction of this need is indirect trough the applications that LTE-A makes available for the user, and in particular Internet Access with web 2.0 capabilities
Working life	Satisfaction of this need is indirect trough the applications that LTE-A makes available for the user, and in particular Internet Access with e-Business
Transactions	Satisfaction of this need is indirect trough the applications that LTE-A makes available for the user, and in particular Internet Access with e-commerce
Entertainment	LTE-A increases data rates and QoS making possible to deliver high quality services to the end-user, such as mobile TV and multimedia services
Security and Privacy	LTE-A is compliant with the need for privacy and security, providing mechanisms and solutions such as strong mutual subscriber/network authentication, user identity protection, strong encryption of user traffic, and strong encryption and integrity protection of signalling
Health and wellness	Satisfaction of this need is indirectly provided by the applications supported by the technology, such as e-health, which however is only in part fully accessible via a mobile phone. More appeal can be seen is the technology as a vehicle to provide advanced applications for home care and independent living

3.1.6 Scientific research potential in the LTE-A context

The potential for scientific research in the framework of LTE-A is considerable. Researchers can concentrate efforts in implementing mechanisms and algorithms able to satisfy the requirements set by the standard. In this sense, research in this field is more evolutionary than revolutionary.

3.2 WiMAX for IMT-Advanced

3.2.1 Brief description of next generation WiMAX technology

In preparation for IMT-Advanced, the IEEE 802.16 Working Group has moved to initiate a new project designated as "802.16m" [8] with the intent of developing enhancements to IEEE 802.16 standard to ensure suitability as an IMT-Advanced proposal. The development of this amendment to IEEE 802.16 addresses "Air Interface for Fixed and Mobile Broadband Wireless Access Systems- Advanced Air Inter-

Table 2. Link of LTE-A technology with ICT trends

ICT trend	*Link with LTE-A technology*
Ideal performance	Trend testified by the introduction of novel algorithms and mechanisms to improve link, network, and system performance
Ubiquity	Trend testified by the high speed mobility requirements that make services and applications available everywhere while on the move
Flexibility	Trend testified by the use of mechanisms to dynamically adapt to channel and network conditions, such as adaptive beamforming, adaptive modulation and coding, dynamic power control range, dynamic and flexible radio resource management, and flexible spectrum usage
Complexity	Trend testified by the attention towards low-cost requirements and power efficiency which is pursued at both network and terminal level
Cognitivity	Trend testified by the presence in LTE-A of the flexible spectrum usage and spectrum sharing technique, strongly related with the Flexibility trend
Opportunism	Trend again in strong link with the Flexibility trend: opportunism is provided by LTE-A with adaptive algorithms to change according to the experience environment, and with the use of diversity via MIMO
Cooperation	Trend testified by the use of beamforming and MIMO techniques
Security	Trend in one-to-one correspondence with the privacy and security need, offering strong mutual subscriber/network authentication, user identity protection, strong encryption of user traffic, and strong encryption and integrity protection of signalling
Miniaturization	Trend present in all technological standard, where the attention is always put in designing smaller terminals and antennas
Convergence	Trend embodied by LTE-A from the application point of view, offering potentials to the quadruple play concept

face", and amends the IEEE802.16 WirelessMAN-OFDMA specification to provide an advanced air interface for operation in licensed bands.

Briefly, the new version of WiMAX shall operate in RF frequencies less than 6 GHz and be deployable in licensed spectrum allocated to the mobile and fixed broadband services, supporting *scalable bandwidths* from 5 to 40 MHz, via a single or multiple RF carriers. 802.16m shall support MIMO, beamforming operation or other advanced antenna techniques. *Peak spectral efficiency* is in compliance with ITU-R requirements, equal to 15 bit/s/Hz in downlink and to 6.75 bit/s/Hz in uplink. IEEE 802.16m shall enable *advanced RRM* for efficient utilization of radio resources, including measurement/reporting, interference management and flexible resource allocation mechanisms. *Security* functions are foreseen which provide the necessary means to achieve protection of the integrity of the system, confidentiality of user-generated traffic and secure access to services.

3.2.2 Individual need satisfaction by WiMAX

The way WiMAX is able to respond to the individual needs identified in Section 1 is schematized in Table 3.

Table 3. Individual needs satisfaction by the WiMAX technology

Individual need	*Degree of satisfaction of the individual needs by WiMAX technology*
Information and learning	As for LTE-A, also WiMAX development intrinsically respond to the requisite for accessing information anywhere and anytime, enabling fruition while on the move and increasing the supported data rate towards the paradigm of broadband for all
Environmental fitness	Supporting mobility up to vey high speed, like LTE-A also WiMAX responds to the need for environmental fitness, allowing the user to receive information and access applications from any geographic point
Social interaction	Satisfaction of this need is indirect trough the applications made available by the standard for the user, and in particular Internet Access with web 2.0 capabilities
Working life	Satisfaction of this need is indirect trough the applications that WiMAX makes available for the user, and in particular Internet Access with e-Business
Transactions	Satisfaction of this need is indirect trough the applications that WiMAX makes available for the user, and in particular Internet Access with e-commerce
Entertainment	WiMAX increases data rates and QoS making possible to deliver high quality services to the end-user, such as mobile TV and multimedia services
Security and Privacy	WiMAX is compliant with the need for privacy and security, providing protection of the integrity of the system, protection and confidentiality of user-generated traffic and user-related data, and secure provisioning and availability of services provided by the system
Health and wellness	Satisfaction of this need is indirectly provided by the applications supported by the technology, such as e-health, which however is only in part fully accessible via a mobile phone. More appeal can be seen is the technology as a vehicle to provide advanced applications for home care and independent living

3.2.3 Mapping WiMAX onto ICT trends

WiMAX is strongly related to the ICT trends described in Section 2 as detailed in Table 4.

3.2.4 Time frame for WiMAX diffusion

Similarly to LTE-A, the next generation WiMAX standard has been submitted to ITU as a response to the call for technologies in the framework of IMT-Advanced, and the time schedule for its diffusion is in compliance with that of ITU-R. The diffusion of the technology on the market is likely to begin its ramp up in 2012, preceding in this way the spread of LTE-A. Its growth and maturity will characterize the following years. It is difficult to forecast that WiMAX may pre-empt LTE growth.

Table 4. Link of WiMAX technology with ICT trends

ICT trend	*Link with WiMAX technology*
Ideal performance	Trend testified by the introduction of novel algorithms and mechanisms to improve link, network, and system performance
Ubiquity	Trend testified by the high speed mobility requirements that make services and applications available everywhere while on the move
Flexibility	Trend testified by the use of mechanisms to dynamically adapt to channel and network conditions, such as adaptive modulation and coding
Complexity	Trend testified by the attention towards low-cost requirements and power efficiency which is pursued at both network and terminal level
Cognitivity	Trend testified by the presence in WiMAX of the self-organization capability for base stations
Opportunism	Trend again in strong link with the Flexibility trend: opportunism is provided by WiMAX with adaptive algorithms to change according to the experience environment, and with the use of diversity via MIMO
Cooperation	Trend testified by the use of beamforming and MIMO techniques
Security	Trend in one-to-one correspondence with the privacy and security need, offering protection of the integrity of the system, protection and confidentiality of user-generated traffic and user-related data, and secure provisioning and availability of services provided by the system
Miniaturization	Trend present in all technological standard, where the attention is always put in designing smaller terminals and antennas
Convergence	Trend embodied by WiMAX from the application point of view, offering potentials to the quadruple play concept. Also, next generation WiMAX is designed to assure maximum compatibility with other technologies

3.2.5 Market impact of WiMAX technology

According to some manufacturers and operators, the market impact of WiMAX in terms of number of subscribers will likely be modest if compared to LTE-A. However, we will have to wait up until 2012 to start drawing definitive conclusions on this.

3.2.6 Scientific research potential in the WiMAX context

Similarly to LTE-A, the scientific potential in the framework of WiMAX is interesting. Researchers can concentrate efforts in implementing mechanisms and algorithms able to satisfy the requirements set by the standard. In this sense, research in this field is more evolutionary than revolutionary.

3.3 MANETs and BANs

3.3.1 Brief description of MANETsand BANs

Identifying a unique standard about MANETs is not possible, since reference to MANETs can be found in many fora, from IETF to IEEE, with the examples of

IEEE 802.11s on mesh networking, IEEE 802.15.3 on high data rate WPANs, IEEE 802.15.4-2006 high data rate WPANs, IEEE 802.15.5, and IEEE 802.15 TG6 which is specifically about BANs.

IETF work on MANETs [9] mainly aims at defining IP routing protocol functionalities suitable for wireless routing applications within both static and dynamic topologies, with increased dynamics due to node motion or other factors. Approaches have to be relatively lightweight in nature, suitable for multiple hardware and wireless environments, with hybrid mesh infrastructures.

IEEE 802.11s [10] is a draft IEEE 802.11 amendment for mesh networking, defining how wireless devices can interconnect to create an ad-hoc network. It extends the IEEE 802.11 MAC standard by defining an architecture and protocol that support both broadcast/multicast and unicast delivery using "radio-aware metrics over self-configuring multi-hop topologies".

IEEE 802.15.3 [11] is the standard for high data rate WPAN (HR-WPAN), able to support rates in the order of 20 Mbit/s and up to 55 Mbit/s. Besides a high data rate, the new standard provides for low power, low cost solutions addressing the needs of portable consumer digital imaging and multimedia applications.

IEEE 802.15.4-2006 [12] is a standard which specifies the physical layer and media access control for low-rate wireless personal area networks (LR-WPANs). The basic framework for 802.15.4 considers a 10-meter communications area with a transfer rate of 250 kbit/s. Tradeoffs are possible to favour more radically embedded devices with even lower power requirements, through the definition of several possible physical layers. Lower transfer rates of 20 kbit/s and 40 kbit/s were initially defined, with the 100 kbit/s rate being added in the current revision. Even lower rates can be considered with the resulting positive effect on power consumption. Important features include real-time suitability by reservation of guaranteed time slots, collision avoidance through CSMA/CA and integrated support for secure communications. Devices also include power management functions such as link quality and energy detection.

Finally, the IEEE 802.15 Task Group #6 (TG6) is developing an international standard for BAN. Different use cases are foreseen with BANs: Body Sensor Networks (BSN), Sports and Fitness Monitoring, Wireless Audio, Mobile Device Integration, and Personal Video Devices; each of these use cases unique requirements in terms of bandwidth, latency, power usage, and signal distance.

3.3.2 Individual need satisfaction by MANETs and BANs

The way MANETs and BANs are able to respond to the individual needs identified in Section 1 is schematized in Table 5.

3.3.3 Mapping MANETs and BANs onto ICT trends

MANETs and BANs are strongly related to the ICT trends described in Section 2 as detailed in Table 6.

Table 5. Individual needs satisfaction by MANETs and BANs technology

Individual need	*Degree of satisfaction of the individual needs by MANETs and BANs technology*
Information and learning	MANETs and BANs development intrinsically respond to the requisite for accessing information anywhere and anytime, enabling fruition while on the move and increasing the supported data rate towards the paradigm of broadband for all
Environmental fitness	Supporting mobility and implementing self-configuration, MANETs and BANs respond to the need for environmental fitness, adapting the network topology and allowing the user to receive information and access applications from any geographic point
Social interaction	Satisfaction of this need is testified by cooperation between terminals often foreseen in ad-hoc networks
Working life	Satisfaction of this need is indirect trough the applications that MANETs and BANs make available for the user, and in particular Internet Access with e-Business
Transactions	Satisfaction of this need is indirect trough the applications that MANETs and BANs make available for the user, and in particular Internet Access with e-commerce
Entertainment	Satisfaction of this need is indirect trough the applications that MANETs and BANs make available for the user, and in particular Internet Access and multimedia content
Security and Privacy	MANETs and BANs are compliant with the need for privacy and security, providing mechanisms for intrusion detection and distributed authentication
Health and wellness	Satisfaction of this need is specifically present in BANs and especially for those networks designed for medical applications

3.3.4 Time frame for MANETs and BANs diffusion

Forecasts for MANETs and BANs put the ramp up period of this technology starting in 2015, with maturity period proceeding until 2025 approximately.

3.3.5 Market impact of MANETs and BANs technology

The market impact of MANETs and BANs will likely be extremely interesting; forecasts count up to millions of subscribers reached during the maturity phase.

3.3.6 Scientific research potential in the MANETs and BANs context

The scientific potential in the framework of MANETs and BANs is very considerable, as many areas are open for definition, and large players are less incumbent on the technology with respect to more structured networks and systems. Activities in the relevant standardization fora (IEEE 802.11s, IEEE 802.15) are currently on-going and there is large space for new contributions.

Table 6. Link of MANETs and BANs technology with ICT trends

ICT trend	*Link with MANETs and BANs technology*
Ideal performance	Trend testified by the introduction of novel algorithms and mechanisms to improve link, network, and system performance
Ubiquity	Trend testified by the mobility requirements that make services and applications available everywhere while on the move
Flexibility	Trend testified by the use of mechanisms to dynamically adapt the network topology according to the experienced traffic and environment. Self-configurability and flexibility are present at various levels, such as dynamic routing or distributed medium access control. Also the geographic expansion of the network is flexible, ranging from local area networks to personal and body area networks
Complexity	Trend testified by the attention towards low-cost requirements and power efficiency which is pursued at both network, terminal and sensor level
Cognitivity	In line with the Flexibility trend, this trend is testified by the presence in MANETs of self-configuration and dynamic adjustments of the network structure according to the environment
Opportunism	Again in strong link with the Flexibility trend, opportunism is provided via self-configuration capabilities and dynamic routing
Cooperation	Trend testified by the presence of mesh topologies in the foreseen network topology
Security	Trend in one-to-one correspondence with the privacy and security need, providing mechanisms for intrusion detection and distributed authentication. This is especially true for BANs in medical contexts
Miniaturization	Trend present in all technological standards, where the attention is always put in designing smaller terminals and antennas. This is extremely true for BANs where sensors are extremely miniaturized according to the implant applications
Convergence	Trend embodied by MANETs and BANs from the routing point of view, offering converged solutions coupling reactive and proactive strategies

3.4 DVB-T2

3.4.1 Brief description

The DVB-T2 standard [13], released in September 2009, has been recently published by the DVB forum, defining a new advanced air interface for the provision of digital video broadcasting via the terrestrial network for fixed and mobile terminals, able to *improve efficiencies* in its use of spectrum of 30–50 % compared to the previous DVB-T standard for provision of Digital Terrestrial Television (DTT).

As its predecessor, the DVB-T2 physical layer is based on OFDM (*orthogonal frequency division multiplex*) modulation, with a large number of sub-carriers delivering a robust signal. Also in common with DVB-T, the new specification offers a range of different modes making it a very flexible standard. LDPC (*Low Density Parity Check*) coding combined with BCH (Bose-Chaudhuri-Hocquengham) coding offers excellent performance in the presence of high noise levels and interference, resulting in a very robust signal. 2 × 1 MISO capabilities are considered to provide transmitter diversity through the Alamouti scheme, which improves coverage in small scale single-frequency networks.

3.4.2 Individual need satisfaction by DVB-T2

The way DVB-T2 is able to respond to the individual needs identified in Section 1 is schematized in Table 7.

Table 7. Individual needs satisfaction by DVB-T2 technology

Individual need	*Degree of satisfaction of the individual needs by DVB-T2 technology*
Information and learning	DVB-T2 responds to the requisite of being informed everywhere and at anytime, because even if explicit requirements are set for fixed reception, also portable and mobile terminal can be supported in the enhanced version answering to the paradigm of broadband for all
Environmental fitness	DVB-T2 responds to the need for environmental fitness allowing extreme flexibility of transmission according to the experienced environment
Social interaction	Satisfaction of this need is trough the applications that DVB-T2 makes available for the user, especially considering interactivity capabilities
Working life	DVB-T2 is not designed explicitly to satisfy to this need, except for specific multimedia contents delivered via the network
Transactions	DVB-T2 is not designed explicitly to satisfy to this need, except for specific multimedia contents delivered via the network together with interactivity
Entertainment	DVB-T2 strongly responds to this need offering HDTV and multimedia applications
Security and Privacy	DVB-T2 is compliant with the need for privacy and security, providing cryptography and password protection mechanisms
Health and wellness	DVB-T2 is not designed explicitly to satisfy to this need, except for specific multimedia contents delivered via the network

3.4.3 Mapping DVB-T2 onto ICT trends

DVB-T2 is strongly related to the ICT trends described in Section 2 as detailed in Table 8.

3.4.4 Time frame for DVB-T2 diffusion

DVB-T2 ramp up is foreseen by 2020 after the full switch off from analogue television to DVB-T. Then, its exponential growth until maturity in indubitably forecast by expert in the following ten years.

Table 8. Link of DVB-T2 technology with ICT trends

ICT trend	*Link with DVB-T2 technology*
Ideal performance	Trend testified by the introduction of novel algorithms and mechanisms to improve link, network, and system performance towards the delivery of HDTV to set-top-boxes
Ubiquity	Trend testified by the large geographical penetration foreseen by the standard, even through local mono-language areas
Flexibility	Trend testified by the use of dynamic and flexible mechanisms at various levels, such as in OFDM guard time selection, modulation and coding schemes, OFDM transmission modes
Complexity	Trend testified by the attention towards low-cost requirements and power efficiency which is pursued at both network and terminal level
Cognitivity	In the same sense as Flexibility
Opportunism	In the same sense as Flexibility, with adaptive algorithms to change according to the experience environment, and with the use of diversity via MIMO
Cooperation	Trend testified by the use of MIMO techniques
Security	Trend in one-to-one correspondence with the privacy and security need
Miniaturization	Trend present in all technological standard, where the attention is always put in designing smaller terminals and antennas
Convergence	Trend embodied by DVB-T2 in a double sense: analogue and digital convergence and interoperability with legacy DTT solutions

3.4.5 Market impact of DVB-T2 technology

The market impact of DVB-T2 is likely to be huge, with billions of set-top-boxes and subscribers.

3.4.6 Scientific research potential in the DVB-T2 context

DVB-T2 is already a standard. In this sense, research in this field can be concentrated in providing guidelines for mechanisms implementation, and is more evolutionary than revolutionary. However, its extension towards mobility can offer an interesting topic for investigation to scientific researchers. In fact, a next generation mobile standard is going to be defined in the framework of the DVB-NGH ad hoc group.

3.5 DVB-SH

3.5.1 Brief description of DVB-SH

DVB-SH is the standard for a transmission system designed to deliver video, audio and data services to vehicles and handheld devices [14]. The key feature of DVB-SH is that it is a hybrid satellite/terrestrial system that allows the use of the satellite means to achieve coverage of large regions or even a whole country. In areas where direct reception of the satellite signal is not possible, and a terrestrial gap filler can be used seamlessly to provide coverage. Operating in S-band, the burden placed on DVB-SH by the high frequency is compensated via a selection of tools that enhance the signal robustness: 3GPP2 Turbo codes, OFDM with different modes, use of scattered pilots, continual pilots, and TPS (Transmitter Parameter Signaling) carriers, mode and stream adaptation.

3.5.2 Individual need satisfaction by DVB-SH

The way DVB-SH is able to respond to the individual needs identified in Section 1 is schematized in Table 9.

3.5.3 Mapping DVB-SH onto ICT trends

DVB-SH is strongly related to the ICT trends described in Section 2 as detailed in Table 10.

3.5.4 Time frame for DVB-SH diffusion

DVB-SH is foreseen to start its ramp up period around 2015 achieving maturity and starting fade out likely in 2022.

3.5.5 Market impact of DVB-SH technology

The market impact of DVB-SH will likely be extremely interesting; forecasts count up to millions of subscribers reached during the maturity phase.

Table 9. Individual needs satisfaction by DVB-SH technology

Individual need	*Degree of satisfaction of the individual needs by DVB-SH technology*
Information and learning	DVB-SH responds to the requisite of being informed everywhere and at anytime, thanks to the mobility supported by the standard
Environmental fitness	DVB-SH responds to the need for environmental fitness allowing extreme flexibility of transmission according to the experienced environment and support high speed mobility
Social interaction	Satisfaction of this need is trough the applications that DVB-SH makes available for the user, especially considering interactivity capabilities
Working life	DVB-SH is not designed explicitly to satisfy to this need, except for specific multimedia contents delivered via the network
Transactions	DVB-SH is not designed explicitly to satisfy to this need, except for specific multimedia contents delivered via the network together with interactivity
Entertainment	DVB-SH strongly responds to this need offering mobile TV and multimedia applications
Security and Privacy	DVB-SH is compliant with the need for privacy and security, providing cryptography and password protection mechanisms
Health and wellness	DVB-SH is not designed explicitly to satisfy to this need, except for specific multimedia contents delivered via the network

Table 10. Link of DVB-SH technology with ICT trends

ICT trend	*Link with DVB-SH technology*
Ideal performance	Trend testified by the introduction of novel algorithms and mechanisms to improve link, network, and system performance towards the delivery of mobile TV to handheld terminals
Ubiquity	Trend testified by the large geographical penetration foreseen by the standard, even through local mono-language areas, eased by the satellite means
Flexibility	Trend testified by the use of dynamic and flexible mechanisms at various levels, such as modulation and coding schemes, transmission modes (TDM vs. OFDM), stream adaptation
Complexity	Trend testified by the attention towards low-cost requirements and power efficiency which is pursued at both network and terminal level
Cognitivity	In the same sense as Flexibility
Opportunism	In the same sense as Flexibility, with adaptive algorithms to change according to the experience environment
Cooperation	Trend testified by the coordinated functionalities of the satellite and terrestrial components foreseen in the network
Security	Trend in one-to-one correspondence with the privacy and security need
Miniaturization	Trend present in all technological standard, where the attention is always put in designing smaller terminals and antennas
Convergence	Trend embodied by DVB-SH in the integration and coordination between satellite and terrestrial technologies

3.5.6 Scientific research potential in the DVB-SH context

DVB-SH is already a standard. In this sense, research in this field can be concentrated in provide guidelines for mechanisms implementation and is more evolutionary than revolutionary. However, a next generation standard is going to be defined in the framework of the DVB-NGH ad hoc group.

3.6 Digital radio

3.6.1 Brief description of digital radio

Recently, the new version of the ITU DRM (Digital Radio Mondiale) system, namely DRM+, is under investigation and testing as the extension of the existing standard to the broadcasting bands up to 174 MHz. DRM+ has a relatively narrow bandwidth and is designed to fit in the FM broadcast-band plan and a frequency grid of 100 kHz [15], exploiting coded OFDM so that mobile reception will be possible both in multipath environment (up to 300 km/h) as well as in the long echoes Single Frequency Networks. High data rates (from 37 to 186 kbit/s) together with a variety of modulation formats and code rates are considered, which allow to achieve flexibility in offering services and content as well as the good trade-off between data-rate and transmission power. Efficient audio coding will be adopted which allow for application of stereo-compatible 5.1 multi-channel signal.

In February 2007, the new and upgraded version DAB standard i.e. DAB+, was released by ETSI [16]. DAB+ is based on the original DAB standard but uses the more efficient audio codec i.e. HE-AAC v2 (also known as AAC+). It provides the same functionality as the original DAB radio services and offers (as it was in the DRM+ case) the Services Following, traffic announcements and the PAD (Programme-Associated Data) multimedia data (like dynamic labels, news headlines, complementary graphics and images etc.). Moreover, stronger error correction coding has been applied (i.e. the Reed-Solomon codes).

T-DMB (Terrestrial Digital Multimedia Broadcasting) is an audio/video standard, at first developed in South Korea. T-DMB uses MPEG-4 Part 10 (also known as H.264) for the video stream and MPEG-4 Part 3 BSAC or HE-AAC V2 for the audio data. As in DAB+ the Reed-Solomon codes are used for forward error correction. In order to mitigate the influence of the channel effects, DMB modem uses OFDM-DQPSK modulation. T-DMB has been adopted by ETSI as a global standard in December 2007. The last release has appeared in April 2009 [17].

3.6.2 Individual need satisfaction by digital radio

Digital radio standards are able to respond to the individual needs identified in Section 1 is schematized in Table 11.

3.6.3 Mapping digital radio onto ICT trends

Digital radio is strongly related to the ICT trends described in Section 2 as detailed in Table 12.

Table 11. Individual needs satisfaction by digital radio technology

Individual need	*Degree of satisfaction of the individual needs by digital radio technology*
Information and learning	Digital radio responds to the requisite of being informed everywhere and at anytime, thanks to the applications provided to subscribers and to the mobility supported by the standard, especially on vehicles
Environmental fitness	Digital radio responds to the need for environmental fitness allowing flexibility in transmission according to the experienced environment and supporting high speed mobility
Social interaction	Satisfaction of this need is trough the applications that digital radio makes available for the user, especially considering interactivity capabilities
Working life	Digital radio is not designed to satisfy to this need
Transactions	Digital radio is not designed to satisfy to this need
Entertainment	Digital radio strongly responds to this need offering high quality audio applications
Security and Privacy	Digital radio is not designed to satisfy to this need
Health and wellness	Digital radio is not designed to satisfy to this need

Table 12. Link of digital radio technology with ICT trends

ICT trend	*Link with digital radio technology*
Ideal performance	Trend testified by the introduction of novel algorithms and mechanisms to improve link, network, and system performance towards the delivery of high quality audio content to mobile terminals
Ubiquity	Trend testified by the huge geographical penetration foreseen by the standard, via a Single Frequency Network, and by the high speed mobility support
Flexibility	Trend testified by the use of a variety of modulation formats and code rates, which allow to achieve flexibility in offering services and content as well as a good trade-off between data-rate and transmission power
Complexity	Trend testified by the attention towards low-cost requirements and power efficiency which is pursued at both network and terminal level
Cognitivity	In the same sense as Flexibility
Opportunism	In the same sense as Flexibility
Cooperation	Not particularly reflected in digital radio
Security	Not particularly reflected in digital radio
Miniaturization	Trend present in all technological standard, where the attention is always put in designing smaller terminals and antennas
Convergence	Not particularly reflected in digital radio

3.6.4 Time frame for digital radio diffusion

Digital radio is expected to have a time frame similar to DVB-SH with ramp up in 2015, followed by maturity in the next years.

3.6.5 Market impact of digital radio technology

The market impact of digital radio will likely be extremely interesting, with the potential of millions of terminals and subscribers reached during the maturity phase, especially in vehicular applications.

3.6.6 Scientific research potential in the digital radio context

The scientific potential in the framework of digital radio is fair, because the DRM+ standard is currently under definition and testing; thus, the time is mature for contributing.

3.7 Cognitive radio

3.7.1 Brief description of Cognitive Radio

The reference in the framework of cognitive radio is certainly provided by the activities carried out within the IEEE Standards Coordinating Committee 41 (SCC41) with the IEEE 1900 standard series [18], aimed at standardizing next generation radio systems and advanced spectrum management. Because active dialogue between different standardization bodies is crucial for effective definition of cognitive radio standard, IEEE SCC41 has initiated cooperation with ITU, FCC, OFCOM (the independent regulatory and competition authority for the UK communications industries), and SDR (Software Defined Radio) forum. In this framework, special attention has been dedicated in WPI.6 to the vision of OFCOM, which has recently explicit interesting trends in the evolution of dynamic spectrum access technologies.

The main use cases of CR are dynamic spectrum assignment, dynamic spectrum sharing, and distributed radio resource usage optimization, each of one refers to an appropriate architecture that enable operation in heterogeneous environments. General requirements are the context-awareness (both at network and terminal level), reconfigurability (both at network and terminal level), compliance with regulation, cooperation towards spectrum management.

3.7.2 Individual need satisfaction by Cognitive Radio

Cognitive Radio responds to the individual needs identified in Section 1 is schematized in Table 13.

Table 13. Individual needs satisfaction by cognitive radio technology

Individual need	*Degree of satisfaction of the individual needs by cognitive radio technology*
Information and learning	Cognitive Radio responds to the requisite for accessing information anywhere and anytime, enabling fruition in any condition and through various air interfaces
Environmental fitness	Adaptation to the surrounding environment is an inherent characteristic of cognitive radio, where the most suitable technology is selected dynamically according to the experienced conditions
Social interaction	Satisfaction of this need is indirect trough the applications that Cognitive Radio makes available for the user, and in particular Internet Access with web 2.0 capabilities
Working life	Satisfaction of this need is indirect trough the applications that Cognitive Radio makes available for the user, and in particular Internet Access with e-Business
Transactions	Satisfaction of this need is indirect trough the applications that Cognitive Radio makes available for the user, and in particular Internet Access with e-commerce
Entertainment	Satisfaction of this need is indirect trough the applications that Cognitive Radio makes available for the user, and in particular Internet Access and multimedia contents
Security and Privacy	Cognitive Radio is compliant with the need for privacy and security, providing mechanisms for authentication, integrity, and authentication
Health and wellness	Satisfaction of this need is indirect trough the applications that Cognitive Radio makes available for the user, and in particular Internet Access and e-health

Table 14. Link of cognitive radio technology with ICT trends

ICT trend	*Link with cognitive radio technology*
Ideal performance	Trend testified by the introduction of novel algorithms and mechanisms to improve network and system performance
Ubiquity	Trend testified by the fact that Cognitive Radio is by definition designed to be pervasive and adapted to any conditions. Ubiquity is also testified by the integration with geo-location information
Flexibility	Trend testified by the use of mechanisms to dynamically adapt spectrum management, via dynamic spectrum assignment, dynamic spectrum sharing, and distributed radio resources
Complexity	Trend testified by the attention towards low-cost requirements and power efficiency which is pursued at both network and terminal level
Cognitivity	Trend totally identified with the Cognitive Radio technology, which foresees full adaptation to the environment, traffic, and network conditions
Opportunism	Trend again in strong link with the Flexibility trend: opportunism is provided by Cognitive Radio via Reconfiguration
Cooperation	Trend testified by the use of spectrum sharing techniques
Security	Trend in one-to-one correspondence with the privacy and security need
Miniaturization	Trend present in all technological standard, where the attention is always put in designing smaller terminals and antennas
Convergence	Trend embodied by Cognitive Radio where convergence in among the different technologies foreseen for reconfiguration. Cognitive Radio can be seen as the convergence between all other technologies: cellular, wireless ad-hoc and broadcasting

3.7.3 Mapping Cognitive Radio onto ICT trends

Cognitive Radio is strongly related to the ICT trends described in Section 2 as detailed in Table 14.

3.7.4 Time frame for Cognitive Radio diffusion

It is likely that Cognitive Radio will start its ramp up period approximately after 2015 with maturity reached in about 2020.

3.7.5 Market impact of Cognitive Radio technology

The market impact of Cognitive Radio is uncertain to predict, especially due to technological constraints, even if in principle Cognitive Radio has the potential to achieve a huge market response.

3.7.6 Scientific research potential in the Cognitive Radio context

The scientific potential in the framework of Cognitive Radio is really dramatic. We can safely state that the topic is really hot in the scientific arena.

3.8 *Future Internet*

3.8.1 Brief description of the Future Internet concept

The Internet of the future is more related to applications than to technology. It comprises:

- participative web, i.e. transformation of the Web from a network of separate applications and content repositories to a seamless and interoperable whole, with Social Network Sites at its basis;
- semantic web, whereby data is enriched by meta-information to allow efficient content search and filtering, in a common framework that allows data to be shared and reused across application, enterprise, and community boundaries, with semantic application platforms and statement-based data-storage (distributed semantic databases);
- intelligent applications, consisting in natural language processing, machine learning, machine reasoning, and autonomous agents that process meaning to enable the semantic web;
- open technologies, i.e. open APIs (Application Programming Interfaces) and protocols, widgets (portable code that any user can install and execute on its webpage), open data formats, open-source software platforms, and open data;
- network computing, i.e. Software-as-a-Service (SaaS) or cloud computing, the shared use of distributed computing resources as an alternative to in-house IT applications using local servers and personal devices (offering scale flexibility and cost-efficiency), which translates into Web services and distributed computing (for example, grid computing is a form of distributed computing whereby a "super and virtual computer" is composed of a cluster of networked, loosely-coupled computers, acting in concert to perform very large tasks);
- portable identity (to allow Internet users to log on to many different web sites using a single digital identity) and roaming of portable identity and personal data;
- ubiquitous connectivity, i.e. broadband adoption and mobile Internet access via mobile and portable devices.

A whole new picture opens up when the main objective becomes the interconnections of objects, in a so-called Internet of Things (IoT). IoT entails seamless and self-configuring connection of devices, sensors, objects, rooms, machines, vehicles, etc. through fixed and wireless networks, based on machine-to-machine communications and RFID smart tags. A deep analysis of technical challenges related to Future Internet can be found in [1].

3.8.2 Individual need satisfaction by Future Internet

The way Future Internet is able to respond to the individual needs identified in Section 1 is schematized in Table 15.

3.8.3 Mapping Future Internet onto ICT trends

Future Internet is strongly related to the ICT trends described in Section 2 as detailed in Table 16.

Table 15. Individual needs satisfaction by future Internet technology

Individual need	*Degree of satisfaction of the individual needs by future Internet technology*
Information and learning	Future Internet provides the platform for applications and data storage for information and learning
Environmental fitness	Future Internet provides the most part of the contents that the user may want to receive while on-the-move and the reference structure for environmental fitness
Social interaction	Future Internet is the central stage for ICT social interaction via the social web (web 2.0 and 3.0)
Working life	Future Internet is a new unavoidable platform for professional telework, via e-Business
Transactions	Future Internet is the new digital enabler for e-commerce transactions
Entertainment	Future Internet provides new means to entertain, such as IPTV, communities, chats, on-line gaming
Security and Privacy	Future Internet embodies the need for security and privacy and offers many tools to assure them
Health and wellness	Future Internet provides the platform and enabling data storage for remote health and wellness applications

Table 16. Link of future Internet technology with ICT trends

ICT trend	*Link with future Internet technology*
Ideal performance	Trend testified by the introduction of novel algorithms and mechanisms to improve network performance and user experience
Ubiquity	Trend testified by the global coverage characterizing the Internet, but also by the new concepts of cloud distributed computing
Flexibility	Trend testified by the dynamic addressing and routing algorithms
Complexity	Trend testified by the attention towards low-cost requirements, simplicity of usage, and energy sustainability algorithms
Cognitivity	Trend testified by the presence in Future Internet of the use of different networks as infrastructure, with seamless continuity between all networks
Opportunism	Trend again in strong link with the Flexibility trend
Cooperation	Trend testified by the operation of nodes with cloud and distributed computing
Security	Trend in one-to-one correspondence with the privacy and security need. Diagnosis tools are provided for problem detection and root cause analysis
Miniaturization	With Future Internet, miniaturization can be seen as miniaturization of servers, databases and memories
Convergence	Trend embodied by the integration and cooperation of the different kinds of Internet: Internet of Services vs. Internet of Things vs. Internet of people

3.8.4 Time frame for Future Internet diffusion

The forecast for Future Internet is to ramp up in 2020, with exponential growth in the following years up to achieving maturity.

3.8.5 Market impact of Future Internet technology

The market impact of Future Internet is foreseen to play a formidable role, with an very huge number of users.

3.8.6 Scientific research potential in the Future Internet context

The scientific potential in the framework of Future Internet is the largest, focused on a longer-term period, and thus offering a great potential also for revolutionary research ideas.

4 Social impact: the Information Divide

The pervasiveness of technology in ICT presents clear benefits in terms of quality of life. However, there are also costs and negative consequences, which cannot and should not be neglected. The engineering point of view must here subside in favour of sociological considerations.

Let us imagine for an instant that the provocative assumption "all of the world's knowledge will be on the Internet by 2020" holds true. Although no one will ever be able to prove this, it is a fact that we are witnessing incredible growth in both the number of new web sites opened daily and in Internet traffic. It is also becoming a habit to verify facts and notions by matching what we know with the results of a web search. It is clear that, under the above assumption, the possibility to access the Internet becomes a major discriminator for individuals, groups, and even entire nations, creating a fundamental issue which we here identify as the *Information Divide*, intended as the separation between one entity's knowledge and the rest of all information.

The information divide is not limited to a lack of technology infrastructures, because it can exist even with full access capabilities. Indeed, the information divide can take on two totally different, or dual, forms: the *digital divide* and the *psychological divide*.

The digital divide, i.e. the lack of proper telecommunication infrastructures, has critical impact both in professional and social terms, which can be unbearable for the younger generations. We can distinguish two kinds of digital divide: geographical and demographic. The geographical divide applies between continents and between specific countries within a continent. The geographical divide especially affects developing countries, where the benefits of the ICT revolution are less clear, to the point where the gap with the developed countries tends to be exacerbated. Focused efforts and investments are needed to transform the divide into a digital opportunity.

The demographic divide refers to the clear difference in quality of service and coverage between urban areas and areas which are less densely populated, broadly defined as rural areas. The problem is one of economy of investment, and it cannot be expected that private entities take on all of the costs associated to giving full connectivity to rural areas. Again, institutional effort is required: bridging the digital divide requires large investments in telecommunications infrastructures, which can range from the cost-effective ADSL solutions to the super-high capacity optical fibers, from broadband cellular systems (which enable high penetration) to broadband satellite links (providing large coverage in remote areas).

The second type of divide is the much less explored *psychological divide*, which is related to the inability to handle information, due to either information overload or illiteracy in new technology matters. The psychological divide is related to age and education, as well as to the inability to handle huge amounts of information.

The generational divide relates to the fact that younger generations are more used to technology with respect to older people. In fact, they are born with technology, and have a hard time imagining a world without it. It is a fact that having children in a household largely increases the demand for having Internet access. We are facing the paradox whereby young unexperienced people understand more of the technological world than older and wiser people, with an impact on familiar and social relationships. The educational divide derives by differences in the scholastic background. Several studies have confirmed that a small percentage of the people with lower secondary education access the Internet, while the percentage rapidly grows to saturation for people with higher and higher education.

Finally, there exists a psychological problem in dealing with instantly available infinite information. While it is possible to download virtually unlimited amounts of data on any subject, there is no doubt that *knowledge without experience is not wisdom*. The progress towards the semantic web is a step towards pre-filtering of information in terms of its relevance, which should help considerably in reducing the information overload. But will this be sufficient? And from the professional point of view, what is the impact of the fact that the know-how will not be a major discriminator in the future? These fundamental questions are destined to remain open until new conceptual exploration methods based on creative thinking will be recognised as a disciplined science.

5 Conclusions

This section contains a view on trends currently characterising the ICT world, with special focus on wireless technologies, based on the activities carried out within WPI.6.

Personal needs are addressed at the beginning, around various areas, i.e.: information and learning, environmental fitness, social interaction, working life, health and wellness, security and privacy, entertainment, and transaction. A brief description is done of each one, followed by a bridging to technologies, services and applications. This listing is intended to provide the rationale for the trends observed in ICT and corresponding technologies growth and evolution.

An overview of ICT trends follows via an original metaphor of the current ICT picture and biological DNA. Two meta-trends are identified, around personalisation and distribution, a discussion on these concepts being provided. Then, ten trends are identified: Ideal performance, Ubiquity, Flexibility, Complexity, Cognitivity, Opportunism, Cooperation, Security, Miniaturisation, and Convergence. An analysis of each one is provided, which includes a mapping onto the two meta-trends.

Cutting-edge standards and technologies in the ICT wireless context are then addressed: LTE-A, WiMAX, MANETs and BANs, DVB-T2, DVB-SH, Digital Radio, Cognitive Radio, and Future Internet. For each one, a mapping onto ICT trends is provided, showing which trends are embodied by the technology and which personal needs can be satisfied in this manner. Along with this analysis, an estimate of the time frame characterising each of these standards is presented, together with a forecast the market impact of each technology, and a measure of the potential for scientific research underneath each standard, with the objective to guide NEWCOM++ researchers towards the most cutting-edge thematic.

References

1. Corazza G., Pedone R., Vanelli-Coralli A.: Technology as a need: Trends in the evolving information society. Advances in Electronics and Telecommunications **1**(1), 124–132 (2010).
2. Maslow A.: Motivation and Personality. 3rd ed. Addison-Wesley, New York NY (1987).
3. i2010 – A European Information Society for growth and employment. [Online]. Available: http://ec.europa.eu/information_society/eeurope/i2010/index_en.htm
4. Karulf E.: Body Area Networks (BAN). [Online]. Available: http://www.cse.wustl.edu/jain/cse574-08/ftp/ban/index.html
5. NEWCOM++ WPI.6, DI.6.1: Report on trends in communication technologies. Tech. Rep. (2008).
6. NEWCOM++ WPI.6, DI.6.2: Report on requirements and constraints in communication technologies. Tech. Rep. (2009).
7. 3GPP Technical Specification Group Radio Access Network. Requirements for further advancements for Evolved Universal Terrestrial Radio Access (E-UTRA) (LTE-Advanced) (Release 8). Tech. Rep. (2009).
8. IEEE 802.16 Broadband Wireless Access Working Group. Work plan for IEEE 802.16m standard & imt-advanced submission. Tech. Rep. (2009).
9. Mobile Ad-hoc Networks (MANET) Working Group. [Online]. Available: www.ietf.org/dyn/wg/charter/manet-charter.html
10. IEEE 802.11 Wireless LAN Working Group. Hwmp protocol specification. Tech. Rep. (2006).
11. Part 15.3: Wireless Medium Access Control (MAC) and Physical Layer (PHY) Specifications for High Rate Wireless Personal Area Networks (WPANs). IEEE Std. 802.15.3-2003 (2003).
12. Part 15.4: Wireless Medium Access Control (MAC) and Physical Layer (PHY) Specifications for Low-Rate Wireless Personal Area Networks (WPANs). IEEE Std. 802.15.4c-2009 (2009).
13. Digital Video Broadcasting (DVB); Frame structure channel coding and modulation for a second generation digital terrestrial television broadcasting system (DVB-T2). ETSI Std. EN 302 755 V1.1.1 (2009-09) (2009).
14. Digital Video Broadcasting (DVB); Framing Structure, channel coding and modulation for Satellite Services to Handheld devices (SH) below 3 GHz. ETSI Std. ETSI EN 302 583 V1.1.1 (2008-03) (2008).
15. DRM+ Presentation. [Online]. Available: http://www.drm.org/drm-the-system/misc/drm-presentation/

16. Digital Audio Broadcasting (DAB); Transport of Advanced Audio Coding (AAC) audio. ETSI Std. ETSI TS 102 563 V1.1.1 (2007-02) (2007).
17. Digital Audio Broadcasting (DAB); DMB video service; User Application Specification. ETSI Std. ETSI TS 102 428 V1.2.1 (2009-04) (2009).
18. IEEE Standard for Architectural Building Blocks Enabling Network-Device Distributed Decision Making for Optimized Radio Resource Usage in Heterogeneous Wireless Access Networks. IEEE Std. 1900.4-2009 (2009).

Heterogeneous and opportunistic wireless networks

Jordi Pérez-Romero (✉), Sergio Palazzo, Laura Galluccio, Giacomo Morabito, Alessandro Leonardi, Carles Antón-Haro, Erdal Arıkan, Gerd Asheid, Marc Belleville, Ioannis Dagres, Davide Dardari, Mischa Dohler, Yonina Eldar, Laura Galluccio, Sinan Gezici, Lorenza Giupponi, Christine Guillemot, George Iosifidis, Michel Kieffer, Marios Kountouris, Alessandro Leonardi, Marco Luise, Guido Masera, Javier Matamoros, Natalia Miliou, Giacomo Morabito, Dominique Morche, Christophe Moy, Carles Navarro, Gabriella Olmo, Sergio Palazzo, Jacques Palicot, Raffaella Pedone, Venkatesh Ramakrishnan, Shlomo Shamai, Piotr Tyczka, Luc Vandendorpe, Alessandro Vanelli-Coralli, Claudio Weidmann, and Andreas Zalonis

Recent years have witnessed the evolution of a large plethora of wireless technologies with different characteristics, as a response of the operators' and users' needs in terms of an efficient and ubiquitous delivery of advanced multimedia services. The wireless segment of network infrastructure has penetrated in our lives, and wireless connectivity has now reached a state where it is considered to be an indispensable service as electricity or water supply. Wireless data networks grow increasingly complex as a multiplicity of wireless information terminals with sophisticated capabilities get embedded in the infrastructure.

When looking at the horizon of the next decades, even more significant changes are expected, bringing the wireless world closer and closer to our daily life, which will definitely pose new challenges in the design of future wireless networks. In the following sections, a vision is briefly described on some of the envisaged elements that will guide this evolution, and that will definitely impact of the design of the network layers of wireless networks.

In particular, and addressing a shorter term perspective, the paper starts in Section 1 with the implications of network heterogeneity on how wireless networks will be operated and perceived by the users, with the corresponding requirements for smart radio resource management strategies. Then, Section 2 will address the concept of self-organising networks, since it is expected that, due to the increase in complexity in future networks, together with the huge amount of parameters to be tuned for network optimisation, mechanisms are established to allow the self-configuration of a wireless network with minimum human intervention. Going a step further from the more classical cellular networks and their evolutions, the next sections address other types of wireless communications networks that are also envisaged to arise

Benedetto S., Correia L.M., Luise M. (eds.): The Newcom++ Vision Book. Perspectives of Research on Wireless Communications in Europe.
DOI 10.1007/978-88-470-1983-6_3, © Springer-Verlag Italia 2012

in the next years. Specifically, Section 3 will be devoted to opportunistic networks, while Section 4 will present the application of wireless technologies in vehicular communication networks and finally Section 5 will be devoted to the Future Internet and the Internet of Things concept, addressing in particular the implications from the wireless perspective. The paper will end in Section 6 with a look towards the long term future by analysing the possibility to extend networking concepts to the nanoscale communications, in the so-called nanonetwork concept, a futuristic interdisciplinary field involving both networking and biological aspects, with still many open research problems but that can generate a new paradigm in how networks can be understood in a time horizon of some decades.

1 Heterogeneous networks and adaptive RRM

A clear trend during the last decade has been a very significant increase in the user demand for wireless communications services, moving from classical voice service towards high bit rate demanding data services, including Internet accessibility everywhere and anytime. While it could have appeared that the penetration of wireless data services experienced a slow launching time, like in the case of Internet access through 2.5G networks such as GPRS, maybe one of the main reasons for that was that the achievable bit rates of only some tenths of kb/s were still very low to support a data service with similar characteristics to that of the wired Internet connections. In fact, this trend was completely changed with the appearance of 3G and particularly 3.5G networks like HSDPA. Thanks to the significantly larger bit rates available, in the order of a few Mb/s, wireless Internet access was much more attractive for final users, with the corresponding explosion in the usage of devices enabling wireless Internet access (e.g. USB modems, modems integrated in laptops, etc.). In that respect, and given that current applications are more and more bit rate demanding, such as multimedia traffic, games, 3D applications, etc., there is no particular reason to envisage that the trend towards demanding higher bit rates in the wireless segment will change in the short future. In fact, this is the trend that can already be appreciated under the development of future LTE and LTE-Advanced systems, which, as a difference from previous generations like GSM or UMTS, now place the focus on the packet data services while the provision of classical voice services is seen as something rather complementary that will be mainly provided through 2G/3G legacy networks.

From a general perspective, the provision of wireless communication services requires from the operator perspective to carry out an appropriate network dimensioning and deployment based on the technologies available and in accordance with the expected traffic demand over a certain geographical area. The target should be then to ensure that services are provided under specific constraints in terms of quality observed by the user, related both to accessibility (e.g. coverage area, reduced blocking/dropping probabilities) and to the specific service requirements (e.g. bit rate, delay, etc.), while at the same time ensuring that radio resources are used efficiently so that operator can maximize capacity. Then, when looking at the basic ingredients to deal with an increasing demand of traffic and bit rate in wireless communication services, this can mainly come through the following general principles:

- *Increasing the spectral efficiency, i.e. the number of bits/s that can be delivered per unit of spectrum.* This can be achieved from the technological perspective, improving the transmission mechanisms at the different layers (e.g. efficient access techniques, efficient radio resource management mechanisms, etc.). This trend has also been observed in the past evolution of mobile technologies with the adoption of more efficient multiple access techniques in each mobile generation, (e.g. from TDMA in GSM to CDMA in UMTS and OFDMA in LTE), with the inclusion of adaptive Radio Resource Management (RRM) mechanisms that try to better exploit the different channel conditions perceived by users (e.g. link adaptation strategies, fast packet scheduling, etc.), and with the adoption of MIMO technologies.
- *Increasing the number of base stations to provide a service in a given area,* thus reducing the coverage area of each one. On the one hand, this allows that all resources available in the base station are used in a smaller area, resulting in a larger efficiency in terms of resources per unit of area. On the other hand, this also allows mobile users being connected closer to their serving base stations, resulting in better channel conditions and consequent larger bit rates. The evolutionary trend in this direction has been in the past the provision of services using hierarchical layers of macro, micro and pico cells, the latter for high traffic density areas, while it is envisaged in the short future the deployment of femto-cells or home base stations that has already started in some places. In fact, some predictions from 2006 in [1] envisaged 102 millions of femtocell users over 32 millions of access points worldwide for 2011.
- *Increasing the total available bandwidth,* i.e. having a larger amount of radio spectrum to deploy the services. In that respect, there has always been a continuous trend to increase the occupied spectrum. Starting from the 50 MHz initially assigned to GSM in the band of 900 MHz, this was increased towards roughly 350 MHz currently occupied by 2G and 3G systems in frequencies of up to 2.1 GHz. Following this expansion, and although there will be still some room for allocating new systems, such as in the case of LTE, the room for improvement here is quite limited, since there is less and less spectrum available in those frequency bands that would be more suitable for deploying a wireless network. To overcome this issue, this will claim for the adoption in the next decades of novel spectrum usage paradigms that enable a more efficient usage of the available spectrum.

In accordance with the above discussions, the natural evolution that one can envisage for the next decade in wireless communications networks resides on the principles of heterogeneity due to the coexistence of multiple technologies with different capabilities and cell ranges, on the need to have smart and efficient strategies to cope with the high demand of broadband services, and also on a longer term basis on the introduction of flexibility in the way how spectrum is being managed. These aspects are further discussed in the following points.

Heterogeneity in networks and devices. Multiple technologies are being deployed one on top of the other, such as the cellular evolution GSM-UMTS-HSPA-LTE or the profusion of WLANs and short-range technologies (e.g. Bluetooth, RF-ID). Then, the future wireless arena is expected to be heterogeneous in nature, with a multiplicity of cellular, local area, metropolitan area and personal area technologies coexisting in the same geographical region. In addition even for a given cellular technology also different deployments will be envisaged particularly in urban areas, where macro, micro, pico and femtocell deployments will be coexisting as the means to achieve the desired large capacities. Heterogeneity will not only be present from the network perspective but also very different types of wireless devices will exist. In addition to classical phones, also PDAs, navigators, sensors or more simple RF-ID devices will constitute different ways of having access to the wireless services.

Network heterogeneity has been in fact regarded as a new challenge to offer services to the users over an efficient and ubiquitous radio access thanks to coordinating the available Radio Access Technologies (RATs), which on the other hand exhibit some degree of complementariness. In this way, not only the user can be served through the RAT that fits better to the terminal capabilities and service requirements, but also a more efficient use of the available radio resources can be achieved. This calls for the introduction of new RRM algorithms operating from a common perspective that take into account the overall amount of resources available in the existing RATs, and therefore are referred to as Common RRM (CRRM), Joint RRM (JRRM) or Multi RRM (MRRM) [2].

In this context, smart RAT selection and vertical handover mechanisms are able to achieve significant capacity gains in networks such as GSM/GPRS/UMTS in which performance obtained can be in a similar order of magnitude [3, 4]. Nevertheless, when considering high bit rate networks such as HSPA and even more LTE or LTE-Advanced, the big difference in terms of performance when compared to e.g. GPRS or UMTS R99, one could expect that capacity improvements achieved thanks to CRRM are only marginal. In spite of this fact, even in this case the complementarities of the access networks can be used to cope with service and terminal differentiation, in the sense that legacy networks can still be used to provide those less bit rate demanding services (e.g. voice) and to provide service to those terminals not yet supporting new technologies. This would allow improving the peak bit rates attained by high demanding services at the new technologies such as LTE.

Multiplicity of data services with high bit rate requirements. We have already witnessed the explosion of wireless data services, mainly thanks to the deployment of wideband technologies (e.g. HSPA) that allow a user experience closer to the one that can be achieved in wired networks. In future years, it is expected that this trend continues and even that other services achieve more significant market-shares, such as mobile TV, multimedia, games, etc., and also some other specific services dealing with healthcare (e.g. through biosensors) or location-based services thanks to the proliferation of navigators.

The convergence between mobile and data access internet-based services posed specific challenges to wireless networks designers about how to exploit the set of resources available as efficiently as possible. In the first wireless networks, since the only application supported was voice, main requirements were to keep a good subjective voice quality and a bounded blocking probability to ensure accessibility in the coverage area. These two aspects were easily handled by means of an adequate network planning that ensured the SINR constraints in the coverage area and that a sufficient number of channels were available. However, with the evolution of wireless networks to support the provision of different types of services with different requirements (e.g. mixing both real time and non-real-time applications with different degrees of user profiles) and the development of more sophisticated and flexible radio access technologies (e.g. based on CDMA or OFDMA), QoS provision cannot be achieved only through a static planning procedure, but it is dynamically pursued by a set of functionalities grouped under the general term "Radio Resource Management" (RRM) [5, 6].

RRM functions are responsible for taking decisions concerning the setting of different parameters influencing on the radio interface behaviour. This includes aspects such as the number of simultaneous users transmitting with their corresponding powers, transmission bit rates, the corresponding code sequences or sub-carriers assigned to them, the number of users that can be admitted in a given cell, etc. Since the different RRM functions will target to track different radio interface elements and effects, they can be classified according to the time scales they use to be activated and executed. In that sense, RRM functions such as power control, scheduling or link adaptation mechanisms tend to operate in short term time scales (typically in the order of milliseconds) while other functions such as admission control, congestion control or handover tend to operate in longer term time scales (typically in the order of seconds).

Among the different RRM strategies, and given the envisaged importance of data services in future networks, it is expected that smart and efficient scheduling mechanisms, aiming at the assignment of resources in the short term to the users requesting service, targeting the maximization of system usage and the fulfilment of QoS constraints, become particularly relevant in the next decade. This has already been the case of 3.5G technologies like HSPA, operating in packet switched mode, so that resources are not permanently assigned to users but they are only assigned when required, and will be the case of future LTE and LTE-Advanced systems.

Trend towards decentralisation and flat architectures. The architecture of the radio access networks has been traditionally hierarchical, composed by a number of base stations connected to some controller node (e.g. the BSC in 2G systems or the RNC in 3G) that is in charge of all the management, decision and control functions related with the radio interface. As a result, RRM functions have been classically implemented in a centralized way, i.e. in central network nodes that can have a more complete picture of the radio access status than a particular node, so that decisions can be made with more inputs. However, centralized implementations have some drawbacks in terms of increased signaling load or transfer delay of the algorithm's inputs to the central node, which prevents an efficient implementation of short-term RRM functions such as packet scheduling, resulting in larger latencies for the information delivered and in a lower capability for the scheduling algorithm to follow channel variations, with the corresponding decrease in spectrum efficiency.

While the above limitations could be acceptable in classical circuit-oriented networks and for low bit rate services, they can bring significant inefficiencies when broadband services are to be provided. This explains why already wireless cellular technology evolutions (e.g. HSPA) exhibit the trend towards implementing RRM functions on the radio access network edge nodes (e.g. base stations), which are in charge of executing fast packet scheduling mechanisms with the corresponding decrease in latencies. To a larger extent, this trend is also present in the SAE (System Architecture Evolution) for LTE, which tries to capture a flat architecture in which decision and control functionalities of the radio interface are moved towards the base stations, including additional capabilities to coordinate among them through direct interfaces (e.g. the X2 interface).

On the other hand when combining this trend with the expected proliferation of femto-cells in future years, which will in fact contribute to an important modification in the classical cellular approaches, it is clear that the degree of coordination among cells should necessarily be limited to relatively slow varying information, otherwise the required signalling to communicate multiple cells would prohibitively increase. The clear result of this will be that autonomous operation of base stations, including the necessary adaptive mechanisms to analyse their environment and take the appropriate configuration decisions, will be a must.

Going one step further in this decentralization direction leads to the implementation of distributed RRM functions at the terminal side, where relevant information for making smarter decisions is kept. This approach has claimed to be inefficient in the past because of the limited information available at the terminal side (e.g. the terminal does not know what is the cell load). Nevertheless, this can be overcome if the network is able to provide some information or guidelines to the terminal assisting its decisions, in addition to the valuable information that is already kept by the terminal (e.g. knowledge of the propagation conditions to surrounding cells). In this way, while a mobile-assisted centralized decision making process requires the inputs from many terminals to a single node, the network-assisted decentralized decision making process requires the input from a single node to the terminals, which can be significantly more efficient from a signalling point of view. An example of this trend is the IEEE P1900.4 protocol.

Flexible spectrum management. The regulatory perspective on how the spectrum should be allocated and utilized in a complex and composite technology scenario is evolving towards a cautious introduction of more flexibility in spectrum management together with economic considerations on spectrum trading. This new spectrum management paradigm is driven by the growing competition for spectrum and the requirement that the spectrum is used more efficiently [7]. Then, instead of the classical fixed spectrum allocation to licensed systems and services, which may become too rigid and inefficient, it is being recently considered the possibility to use Flexible Spectrum Management (FSM) strategies that dynamically assign spectrum bands in accordance with the specific traffic needs in each area [8]. There are different FSM scenarios presenting different characteristics in terms of technical, regulatory and business feasibility. While a fully enabled FSM scenario can be envisaged at a rather long-term perspective (e.g. maybe in more than 20 years), there are already some basic FSM scenarios that will likely become a reality in the short term. In particular, spectrum refarming, providing the possibility to set-up communication on a specific RAT in different frequency bands (e.g. refarming of GSM spectrum for UMTS/HSxPA communications), is a first example. Another case for FSM arises from the so-called digital dividend, which corresponds to the frequencies in the UHF band that will be cleared by the transition of analog to digital television. The cleared spectrum could be utilized by mobile TV or cellular technologies like UMTS, LTE, WiMAX, etc. and also for flexibly sharing spectrum between smart radio technologies. The exploitation of the so-called TV White Spaces, which refer to portions of spectrum that are unused either because there is currently no license holder for them, or because they are deliberately left unused as guard bands between the different TV channels, is another opportunity for FSM mechanisms.

Based on the above trend, a possible classification of spectrum usage models is given in [9] as follows:

- *Exclusive Usage Rights model.* This model consists in the assignment of exclusive licenses under which technology, services and business models are at the discretion of the licensee. In this case, harmful interference is avoided through technology-neutral restrictions. Licenses may be traded in an open market, reducing barriers to entry and encouraging an efficient use of the spectrum.
- *Spectrum Commons model.* This model promotes shared access to available spectrum for a number of users, so that in this case the networks themselves are responsible for ensuring that no interference exists. Under this model, two different levels exist. The first one is the public commons model, in which spectrum is open to anyone for access with equal rights, as it would be the case of e.g. current wireless standards in license-free ISM band. The second possibility is the so-called private commons model, in which the license holder enables multiple parties to access the spectrum in existing licensed bands, under specific access rules set by the primary license holder.
- *Opportunistic spectrum access model.* Under this model, secondary users are allowed to independently identify unused spectrum bands at a given time and place and utilize them while not generating harmful interference to primary license holders. Opportunistic spectrum access may take the form of either underlay access, in which signal powers below the noise floor (e.g. UWB) are used, or overlay access, which involves the detection and use of spectrum white spaces. In this case, cognitive radio networks become key in providing the awareness and adaptability required.

The concept of flexible spectrum management is closely related with the reconfigurability needed in radio equipment in order to cope with variations in the bands allocated to each system. Such reconfigurability can be achieved through the application of SDR technologies. In that respect, different initiatives have been started, such as the ETSI Reconfigurable Radio Systems (RRS) Technical Committee (TC).

It carries out a feasibility study of the standardization activities related to Reconfigurable Radio Systems encompassing radio solutions related to Software Defined Radio (SDR) and Cognitive Radio (CR) research topics, trying to identify requirements from relevant stakeholders, such as manufacturers or operators. In particular, requirements and reference architectures for reconfigurable base stations and mobile devices have been identified, and also a functional architecture for the management and control of RRSs has been developed, including Dynamic Self-Organizing Planning and Management, Dynamic Spectrum Management and Joint Radio Resource Management functionalities [10]. Based on this it is expected for the next decade that reconfigurability will also play a relevant role in wireless communications networks.

2 Self Organising Networks

Third-generation (3G) mobile communication systems have seen widespread deployment around the world. The uptake in HSPA subscriptions, numbering well over 85 million at the start of 2009 [11] indicates that the global thirst for mobile data has just begun: broadband subscriptions are expected to reach 3.4 billion by 2014 and about 80 percent of these consumers will use mobile broadband.

Operators are doing business in an increasingly competitive environment, competing not only with other operators, but also with new players and new business models. In this context, any vision for technology evolution should encompass an overall cost-efficient end-to-end low latency proposition. By the end of 2004, 3GPP started discussions on 3G evolution towards a mobile communication system that could take the telecoms industry into the 2020s. Since then, the 3GPP has been working on two approaches for 3G evolution: LTE and HSPA+. The former is being designed without backward compatibility constraints, whereas the later will be able to add new technical features preserving WCDMA/HSPA investments.

The introduction of LTE and LTE-Advanced systems in the near future over current existing networks will bring a high degree of heterogeneity with many operators having several mobile network technology generations coexisting (GSM/GPRS, UMTS/HSPA, LTE/LTE-Advanced). This will require spending considerable effort in planning, configuring, optimising, and maintaining their networks, consuming significant operational expenditure (OPEX), because with the increasing complexity in the network deployments, network optimization will involve the tuning of a multiplicity of parameters that have a relevant impact over system performance. Traditionally, this tuning was done manually or semi-automatically during initial network deployment phase, leading to static parameter settings. However, being wireless networks inherently dynamic and very sensitive to traffic and interference variations, which is particularly more relevant with the CDMA and OFDMA-based radio interfaces of 3G and LTE systems, this approach can easily lead to inefficient operations, which will become more critical when broadband services are to be provided.

On the other hand, the envisaged high density of small sites, e.g. with the introduction of femto-cells, and the pressure to reduce costs clearly indicate that deploying and running networks needs to be more cost-effective. Consequently, it will be very important for operators that future wireless communication systems bring significant reductions of the manual effort in the deployment, configuration and optimisation activities. Based on this, besides the evolution in radio technologies and network architectures, there is also an evolution in the conception on how these networks should be operated and managed, towards the inclusion of automated Self-Organizing Network (SON) mechanisms to provide the networks with cognitive capabilities enabling its self-configuration without the need of human intervention.

The introduction of Self-Organising Networks (SON) functionalities, aiming to configure and optimize the network automatically, is seen as one of the promising areas for an operator to save operational expenditures and in fact this concept is already being introduced in the context of E-UTRAN standardisation, since it will be required to make the business case for LTE more attractive. It can be envisaged that in the next decades different networks will make use of this concept. Standardization efforts are demanded to define the necessary measurements, procedures and open interfaces to support better operability under multi vendor environment.

As a result of the above, SON has received a lot of attention in recent years in 3GPP [12] or in other initiatives such as the NGMN (Next Generation Mobile Networks) project [13], an initiative by a group of leading mobile operators to provide a vision of technology evolution beyond 3G for the competitive delivery of broadband wireless services to increase further end-customer benefits. It can be consequently envisioned that SON mechanisms will play relevant role in the mobile networks in the framework of the next decade.

A Self-organizing Network (SON) is defined in [14] as a communication network which supports self-x functionalities, enabling the automation of operational tasks, and thus minimizing human intervention. Self-organisation functionalities should be able not only to reduce the manual effort involved in network management, but also to enhance the performance of the wireless network. SON functionalities are classified in [14] in the following categories or tasks:

- *Self-configuration.* It is the process where newly deployed nodes are configured by automatic installation procedures to get the necessary basic configuration for system operation. This process is performed in pre-operational state where the RF interface is not commercially active. After the first initial configuration, the node shall have sufficient connectivity towards the network to obtain possible additional configuration parameters or software updates to get into full operation.
- *Self-planning.* It can be seen as a particular situation of self-configuration comprising the processes were radio planning parameters (such as neighbor cell relations, maximum power values, handover parameters, etc.) are assigned to a newly deployed network node.
- *Self-Optimization.* It is the process where different measurements taken by terminals and base stations are used to auto-tune the network targeting an optimal behavior by changing different operational parameters. This can provide quality and performance improvements, failure reductions and operational effort minimisation. This process is accomplished in the operational state where the RF interface is commercially active.

- *Self-managing.* This corresponds to the automation of Operation and Maintenance (O&M) tasks and workflows, i.e. shifting them from human operators to the mobile networks and their network elements. The burden of management would rest on a mobile network itself while human operators would only have to provide high level guidance to O&M. The network and its elements can automatically take actions based on the available information and the knowledge about what is happening in the environment, while policies and objectives govern the network O&M system.
- *Self-healing.* This process intends to automatically detect problems and to solve or mitigate them avoiding impact to the users. For each detected fault, appropriate alarms shall be generated by the faulty network entity. Self-healing functionality monitors the alarms, and when it finds alarms that could be solved automatically, it analyses more necessary correlated information (e.g. measurements, testing results, etc) to trigger the appropriate recovery actions to solve the fault automatically.

Operator use cases have been formulated for different stages in [15] such as planning (e.g. eNode B power settings), deployment (e.g. transport parameter set-up), optimization (e.g. radio parameter optimization that can be capacity, coverage or performance driven) and maintenance (e.g. software upgrade). Similarly, in [16] different use cases are identified addressing self-optimisation, self-healing and self-configuration functions. These use cases include self-optimisation of Home eNode B, load balancing, interference coordination, packet scheduling, handover, admission and congestion control, coverage hole detection and compensation, cell outage management, management of relays and repeaters and automatic generation of default parameters for the insertion of network elements.

3GPP progress in the context of SON is being reflected e.g. in [17], where use cases are being developed together with required functionalities, evaluation criteria and expected results, impacted specifications and interfaces, etc. Similarly, ICT-FP7 SOCRATES project is developing in deeper detail different use cases, such as cell outage [18]. Paradigms related to SON are discussed in an early paper in this area [19]. Additionally, in [20] the self-configuration of newly added eNode-Bs in LTE networks is discussed and a self-optimisation load balancing mechanism is proposed as well.

Generally self-x functionalities will be based on a loop (self-x cycle) of gathering input data, processing these data and deriving optimized parameters. A generic SON procedure is illustrated in Fig. 1, where network inputs coming from different sources (e.g. network counters, measurement reports, etc.) can be considered in the form of Key Performance Indicators (KPIs). The procedure acts as a loop that continuously interacts with the real network based on *observation* and *actions.* At the observation stage, the optimisation procedure collects information from the different KPIs available. Then, the optimisation procedure will contain different processes for each of the optimisation targets specified based on operator policies (e.g. avoidance of coverage holes, interference reduction, reduction of overlapping between cells, etc.). For each optimisation target a procedure involving the processing the KPIs coming from the observation phase (e.g. combining several KPIs, obtaining

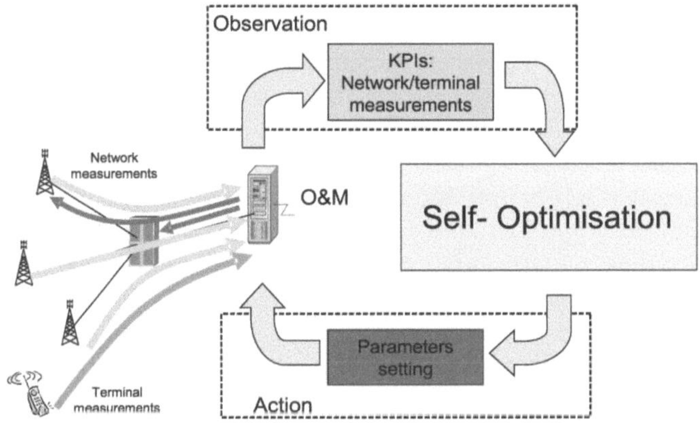

Fig. 1. Network optimisation loop

statistical measurements such as averages or percentiles, etc.), the detection of the sub-optimal operation based on some criteria dependent on the specific target, and the optimisation search to find the proper parameter setting to solve the sub-optimal operation. The final result of the optimisation will be the adequate configuration of selected network parameters (e.g. antenna downtilts, pilot powers, RRM parameters configuration, etc.) affecting either the cell where the sub-optimal operation has been found and/or its neighbouring cells. This change in the configuration parameter(s) turns out to be the *action* executed over the network.

Clearly, the SON concept at the largest extent of a totally automatic network, able to operate with minimum human intervention, is quite ambitious and challenging, so that it can be anticipated that SON will continue as a hot research topic in coming years requiring further research efforts to facilitate its practical implementation. In particular, some of the aspects that must be properly covered include: (1) to find an appropriate trade-off between performance gains and implementation complexity in terms of signalling, computational requirements and measurements, (2) ensure the convergence of the algorithms towards stable solutions within a given timing requirement and (3) ensure the robustness of the mechanisms in front of missing, wrong or corrupted input measurements.

The above context outlines a roadmap for the next decade from current deployed networks (2G, 3G HSPA), mainly managed by centralized remote operations and maintenance (O&M) applications with intensive human intervention, to future SONs (LTE, HSPA+). Targeting a fully SON as an ultimate objective, an evolutionary process can be envisaged, with the progressive inclusion of the SON culture [21]. Given that changing the network configuration/parameterisation is a critical action from the operator's perspective, automated optimisation procedures are likely to be introduced in a step-by-step approach, starting with automatic sub-optimal behaviour detection mechanisms and establishing the corrective actions to overcome them through the semi-automated optimisation of the different parameters, perhaps with some human intervention. An ultimate view, where a joint self-optimisation of several targets is performed will still require a lot of research efforts, due to the complex interactions between these targets. Supporting tools and/or models for this stage will also be needed to estimate the impact of a given parameter change on the live network. In accordance with this envisaged evolutionary approach, in [21], a roadmap is presented starting from the practical application of UMTS optimisation towards automated self-optimisation procedures in LTE.

While SON concepts can be regarded as a relatively short term vision for the next decade, going one step further brings the idea of automated network operation towards a more complete vision envisaging end-to-end goals. This can be framed under what has been coined under the term "cognitive network". In fact, the word 'cognitive' has recently been used in different contexts related to computing, communications and networking technologies, including the cognitive radio terminology coined by. J. Mitola III [22] and the cognitive network concept. Several definitions of cognitive networks have been proposed up to date, but all of them have one thing in common: they present a vision of a network which has the ability to think, learn and remember in order to adapt in response to conditions or events based on the reasoning and prior acquired knowledge, with the objective of achieving some end-to-end goals [23].

The central mechanism of the cognitive network is the cognitive process, which can perceive current network conditions, and then plan, decide and act on those conditions. Then, this is the process that does the learning and decides on the appropriate response to observed network behaviour [24]. The cognitive process acts following a feedback loop in which past interactions with the environment guide current and future interactions. This is commonly called the OODA (Observe, Orient, Decide and Act) loop, which has been used in very different applications ranging from business management to artificial intelligence, and which is analogous to the cognition cycle described by Mitola in the context of cognitive radios.

In the cognitive behaviour, the perception of the stimulus is the means to react to the environment and to learn the rules that permit to adapt to this environment. This is directly related with the observation phase of the cognition cycle, which tries to capture the network status. This involves a large number of measurements and metrics that can be obtained at different network elements. It is important to identify which are the most relevant ones for each phase of the cognition cycle, since measurements relevant for a particular function need to reach the network element where the corresponding function is implemented. Measurements and metrics of interest may be at connection level (e.g. path loss from terminal to cell site, average bit rate achieved over a certain period of time, etc.) or at system level (e.g. cell load, average cell throughput achieved over a certain period of time, etc.). It is worth mentioning here that the observation function may bring the "sensing concept" to all the layers of the protocol stack, depending on which are the aspects to be observed (e.g. not only the measurement of radio electrical signals at the physical layer should be addressed but also the configuration of different applications or network protocols can be also considered). Any means that permits to analyze the environment, and that may be helpful for the adaptation of the communication system to the constraints imposed by the environment, is worth being taken into account.

Being the learning one the most relevant aspects for a cognitive network, many strategies have been envisaged in the literature as learning procedures with the ultimate goal of acquiring knowledge. In particular, machine learning has been widely considered as a particularly suited framework, with multiple possible approaches. The choice of the proper machine learning algorithm depends on the desired network goals. In any case, one of the main challenges here is that the process needs to be able to learn or converge to a solution faster than the network status changes, and then re-learn and re-converge to new solutions when the status changes again, so that convergence issues are of particular importance. Machine learning algorithms can also be organized in different categories (supervised learning, unsupervised learning, semi-supervised learning, and reinforcement learning being the most common ones) depending on how their desired outcome is obtained. Some of the identified machine learning strategies studied in the literature are neural networks, genetic algorithms, expert systems, case-based reasoning, Bayesian statistics, Q-learning, etc.

3 Opportunistic networks

Opportunistic networking primarily stems from mobile ad hoc networking, that is, the research area that applies to scenarios where the complete absence of any supporting infrastructure is assumed. Nowadays there is an ongoing intense work on this topic, however only a few killer applications have been envisaged so far. Opportunistic networks will spontaneously emerge when users carrying mobile devices, such as PDA, laptops, mobile phones, etc., meet. According to the opportunistic networking paradigm, partitions, network disconnections and nodes mobility are more regarded as challenging chances rather than limiting factors or exceptions. Mobility, for example, is considered as a way to connect disconnected network portions; partitions or network disconnections are not regarded as limitations since, in a longer temporal scale and by exploiting a store-and-forward approach, the information will finally flow from a source to the final destination. So in opportunistic networks, delivering can be considered just a matter of time. Opportunistic networking shares also many aspects with delay-tolerant networking, which is a paradigm exploiting occasional communication opportunities between moving devices. These occasional inter-contacts can be either scheduled or random, although conventional delay-tolerant networks (DTNs) typically consider the communication opportunities as scheduled in time. Opportunistic networking and delay tolerant networking have been somehow distinguished so far in the literature, although a clear classification does not exist. However, as a common understanding, opportunistic networking could be regarded as a generalization of delay-tolerant networking with no a-priori consideration of possible network disconnections or partitions. This basically will result in a different routing mechanisms being employed by the two networking paradigms: DTNs will exploit the same Internet routing approach by only taking into account the possibility to have long delays; on the contrary, in opportunistic networks paths will be dynamically built so as to overcome possible disconnections met along the path.

In the next years we expect diffusion of numerous opportunistic networking scenarios where individuals, equipped with very powerful mobile communication devices, like smart phones or PDAs, will establish transparent direct communication links to each other. The technology for supporting such a kind of immersive communications is already available and it is predictable that, although wireless enhanced communications between devices will not replace standard communications such as SMS or phone calls, they will keep increasing their weight in next generation scenarios.

These enhanced networks formed by millions of mobile devices carried out by moving users are expected to increase in size becoming very crowded, especially in case of metropolitan areas. These networks, according to the traditional opportunistic perspective, will be mainly characterized by sparse topology, lack of full connectivity and sporadic communication opportunities. Overlapping between these network communities will be envisaged as the possibility to exploit superposition to strengthen and enhance the security of communication among network members. To this purpose the concept of *community profile* and/or *homophily* can be employed. The community profile will be used to identify the overlapping communities a user belongs to, and the strength of the logical relationships that a user has with other community users. On the other hand, homophily can help to characterize the degree of similarities between users of different communities and the degree of trust between them. This trust could be used in selecting possible forwarder nodes for information according to the opportunistic networking paradigm.

In this context, exploiting the so-called "social-awareness" (i.e., information about the social relationships and interactions among the members) has been recently proposed to perform data forwarding in opportunistic networks. To this purpose, some solutions have been envisaged, which use information on how users join together in social communities, to create efficient content replication schemes in opportunistic networks [25].

Another interesting application envisaged for wireless opportunistic scenarios is related to delay tolerant data broadcasting [26]. More in depth, due to the high density of mobile devices which are expected to be available in short time, these nodes could be exploited as data broadcaster, for example as an alternative to terrestrial and satellite broadcasting systems. This would represent an un-licensed public broadcasting tool alternative to those traditionally used which are also limited by frequency spectrum licence availability and concession rights, as well as censorship or political constraints. Many channels will be available and groups of users willing to share a special type of contents, e.g. music contents, will use the same channel. Epidemic and more controlled approaches for data forwarding will be supported. To allow this contents' dissemination, dock places will be available where contents could be downloaded from the Internet. Podcasting could for example represent an interesting tool for such a distribution of contents. Podcasting [27] is typically thought for downloading contents to a mobile media player. These contents can then be played back on the move. However, the main limitation with the current idea of podcasting is the impossibility of separating move and downloading of contents from a dock media station. This obviously poses limits in terms of use of the service since there could be hours between downloading opportunities. To improve this aspect and broaden the idea of podcasting, the high penetration and density of short-range wireless communications could be exploited. In fact, iPods, Zunes, or smart phones could represent a vehicle of epidemic multihop resource distribution, without relying on any infrastructure support. This podcasting service would allow the distribution of contents using opportunistic contacts whenever podcasting devices are in wireless communication range without any limitation of proximity to dock stations. Such epidemic dissemination approach is expected to be very successful and penetrating especially in urban areas where people meet in office, elevators, public transportations, sport areas, theatres, music events, etc.

All these elements will also contribute to the realization of a *Minority Report*-like communication scenario where information about user profiles and preferences will be stored by the environment at servers and used to foster service personalization and commercial advertisement based on node position and/or habits and requirements. This will obviously open new perspectives for personalized advertisement and marketing which will clearly pass through considerations of user preferences and habits being published through a context profile stored at the dock stations.

Opportunistic networking will be exploited also in the view of distributing contents retrieved by an integrated set of sensors deployed in devices carried by users moving around an interest area. Academic Research Projects proposing similar ideas have recently appeared (e.g. Cenceme [28], Metrosense [29], CRAWDAD [30], etc.) but it is envisaged that in a more large scale sensor network, these contents could be distributed around and infostations used to collect them. This would allow to both derive some detailed and realistic user mobility models, and to characterize variations in interest parameters (for example temperature or humidity) on a metropolitan area scale, as well as to study the statistics of interactions among moving users, even to use them as feedback for studying evolution of diseases propagation, thus trying to prevent them. Also this information about profiles is thought to be integrated into the virtual world of *Second Life*-like environments.

Finally, deep space communications are another significant application scenario especially for delay tolerant and, thus, opportunistic networking too. In fact, in case of satellite communications and/or communications between the international space station and users on the Earth or artificial satellites in the space, incredibly high delay should be considered. This impacts on the possibility to exchange data since often satellites are in visibility only for few minutes every week/month and this short time should be used as much efficiently as possible for transferring data in spite of the high delay (many minutes) and very high bit error rate. In the next years it is foreseeable that deep space communications will have a significant boost. This is because many international space projects have been launched by NASA, ASI, and others in the view of supporting, in 20–30 years, migration of Earth inhabitants to planets and satellites or international space stations. In this case, reliable communications between these people and those on the Earth should be granted and, consequently, opportunistic and delay tolerant networking become a promising and crucial tool to be exploited in such a science-fiction scenario.

4 Vehicular communication networks

Vehicular networks, enabling the wireless communication between different cars and vehicles in a transportation system have been another topic that has received attention during the last years, driven mainly by the automotive industry and also by public transport authorities, pursuing the increase of both safety and efficiency in transportation means. Although a lot of effort has been devoted, no solutions are already available to the mass markets allowing the automatic operation of cars and the communications among them. It is thus expected that the evolution of wireless communications in different aspects, such as adhoc and sensor networks, distributed systems, combined operation of infrastructure and infrastructureless networks, etc. can become an important step so that vehicular communications become a reality in the next decades.

From a more general perspective, the term Intelligent Transportation Systems (ITS) has been used referring to the inclusion of communication technologies to transport infrastructure and vehicles targeting a better efficiency and safety of road transport. Vehicular Networks, also known as Vehicular Ad-hoc NETworks (VANETs), are one of the key elements of ITS enabling the one hand the inter-vehicle communication and on the other hand the communication of vehicles with roadside base station. This communication is intended to exchange different types of messages, such as safety warnings or traffic information, which can help in avoiding both accidents and traffic congestions through appropriate cooperation.

The provision of timely information to drivers and concerned authorities should contribute to safer and more efficient roads, and in fact safety has been the main motivation for the promotion of vehicular networks. It is mentioned in [31] that if preventive measures are not taken, road death is likely to become the third-leading cause of death in 2020. On the other hand, most of the road accidents are associated with specific operations such as roadway departures or collisions at intersections that could be easily avoided by deploying adequate communication mechanisms between vehicles in which they could warn in advance other vehicles about their intentions of departing or arrivals at intersections.

Different initiatives have been launched in the last years in the area of vehicular communications and ITS, involving governmental agencies, car manufacturers and communication corporations. In the United States, the U.S. Department of Transportation launched the Vehicle Infrastructure Integration (VII) and the IntelliDrive programs targeting vehicle to vehicle and vehicle to roadside stations communications through Dedicated Short Range Communications (DSRC) [32, 33]. In the context of ETSI, the Technical Committee on Intelligent Transport Systems (TC ITS) targets the creation and maintenance of standards and specifications for the use of information and communications technologies in future transport systems in Europe [34]. Most of the technical committee's ongoing standardization activities are focused on wireless communications for vehicle-to-vehicle and vehicle-to-roadside communications, with the goal to address the reduction of road deaths and injuries and to increase the traffic efficiency for transport systems with reduction of transport time, contributing to decrease polluting emissions.

Different applications can be envisaged in the context of vehicular communication networks, ranging from safety aspects to a fully automated control of the car that could potentially be driven without human intervention. From the safety point of view, they can be used by vehicles to communicate imminent dangers to others (e.g. detecting an obstacle in the road, informing of an accident) or to prevent others from specific movements such as entering intersections, departing highways, lane changes, etc. and also the communications can help in automatically keeping a safe distance between a vehicle and the car ahead of it, etc. Another application could be related with traffic management, by providing traffic authorities with proper monitoring statistics that could lead to changes in traffic rules such as variable speed limits, adaptation of traffic lights, etc., in order to increase the fluency of traffic in a given road. Similarly, vehicular communications can also assist drivers providing information to facilitate the control of the vehicle, e.g. lane keeping assistance, parking, cruise control, forward/backward collision warnings, providing travel-related information such as maps or gas station locations, etc. Other applications could also be related with enabling faster toll or parking payments or with road surveillance mechanisms [35]. Finally, also private data communications, enabling e.g. Internet access, games, etc. to passengers inside the car can also be envisaged as an application of future vehicular communications.

From a standardisation point of view, the 5.8 – GHz frequency band was allocated by CEPT/ECC for vehicular communications, and an ITS European Profile Standard is being developed based on IEEE 802.11 and constitutes the basis for developing specifications on interoperability and conformance testings [36].

The IEEE 802.11p system, usually referred to as Wireless Access in Vehicular Environments (WAVE) [37], adapts the IEEE 802.11a standard used in WLANs to the vehicular environment, by addressing the specific needs of vehicular networks. Similarly IEEE 1609.X family of standards for WAVE defines the architecture, communications model, management structure, security mechanisms and physical access for high speed (i.e. up to 27 Mbit/s) short range (i.e. up to 1000 m) low latency wireless communications in the vehicular environment [38]. They deal with the management and security of the networks, addressing resource management and network and application layer aspects. In these standards, vehicular networks should be composed by three types of nodes. The first two are vehicles and roadside stations, including, respectively, vehicle On-Board Units (OBU) and RoadSide Units (RSU), the latter being some equivalent to WiFi access points to provide communication with the infrastructure. In addition, a third type of node is the Public Safety On Board Unit (PSOBU) which is a RSU with special capabilities used mainly in special cars such as police, ambulances or fire trucks in emergency situations. The network should support both private data communications (e.g. Internet connectivity on the move) and public communications, with a higher priority for the latter, since they include safety aspects. Although 802.11p and 1609 draft standards specify baselines for developing vehicular networks, many issues are not addressed yet and more research is required.

A widespread adoption of wireless vehicular communication technologies will require efficient use of the radio channel resources, due to the decentralized nature of the networks and the strict quality of service requirements of applications involving traffic safety issues. At the same time, the network infrastructure needs to be designed and implemented to provide reliable network access to objects travelling at high speeds. This will involve specific RRM mechanisms adapted to the characteristics of vehicular networks [39,40].

5 Future Internet and the Internet of Things

Internet, and the possibility to connect any computer with any other around the world constituted in the last decades one of the major cornerstones of telecommunications, turning into one of these revolutionary changes that highly influenced the life and working style of people. Clearly, in this Internet revolution, also wireless technologies played a major role, enabling the connectivity anywhere and anytime without the need of having wired connections. Under this context, the evolution of Internet for the next decade has been coined under the term *Future Internet*, embracing new (r)evolutionary trends in terms of e.g. security, connectivity and context-aware applications, in which again also wireless technologies will become an important and relevant element for the success of the different initiatives.

In particular, one of the envisaged challenging goals for Future Internet is the possibility to interconnect not only people through computer machines (i.e. people connected everywhere and anytime) but also all type of inanimated objects in a communication network (i.e. connecting everything with everything), constituting what has been coined as the *Internet of Things* [41]. It presents a vision in which the use of electronic tags and sensors will serve to extend the communication and monitoring potential of the network of networks. This concept envisages a world in which billions of quotidian objects will report their location, identity, and history over wireless connections, so in order to make it true, this will require dramatic changes in systems, architectures, and communications [42,43].

Different wireless technologies and research disciplines can be embraced under the Internet of Things concept. Although Radio Frequency IDentification (RFID) and short-range wireless communications technologies have already laid the foundation for this concept, further research and development is needed to enable such a pervasive networking world, with the necessary levels of flexibility, adaptivity and security. Such technologies are needed as they provide a cost-effective way of object identification and tracking, which becomes crucial when trying to connect the envisaged huge amounts of devices. Also real-time localization technologies and sensor/actuator networks will become relevant elements of this new vision, since they can be used to detect changes in the physical status of the different things and in their environment. The different things will be enabled with the necessary artificial intelligence mechanisms allowing them to interact with their environment, detecting changes and processing the information to even take appropriate reconfiguration decisions. Similarly, advances in nanotechnology enabling the manipulation of matter at the molecular level, so that smaller and smaller things will have the ability to connect and interact, will also serve to further accelerate these developments.

Obviously, in order to allow things to communicate with each other we need to address them. The problem is defining the most appropriate addressing schemes taking into account the existing standards. In fact, communications in wireless sensor networks will comply with the IEEE 802.15.4 standard, whereas, RFID communications are regulated by the EPCglobal standards. Recently, 6LowPan has been proposed by IETF to support IPv6 over IEEE 802.15.4 communication devices. In fact, IPv6 uses a 128 bit IP address which is large enough to accommodate all imaginable communication devices (consider that IPv4 addresses are almost exhausted). We may think to use IPv6 addresses for RFID tags as well. This will obviously require solutions for the mapping between IPv6 addresses and EPCglobal tag identifiers.

Wireless technologies under the Internet of Things concept will not only have to deal with short range communications but also it can be expected that wide area technologies (e.g. cellular) will also play an important role, since they will provide the different things with the ability to communicate at longer distances. This could be useful e.g. when a user could try to configure remotely through its mobile phone the adequate settings of electronic equipment at home. Appropriate traffic characterisation of these new applications would also be needed in order to adequately perform the network deployment.

Based on all the above, one can envisage for the next decades a new Internet concept, in which humans will be surrounded by a fully interactive and responsive network environment, constituted by *smart things* able to take appropriate decisions in accordance with environment variations and human needs, contributing to increasing the quality of life. Multiple applications can be envisaged, such as the intelligent electronic control of different machines and gadgets at home or in-car, able to configure and personalise in accordance to used preferences aspects as temperature, humidity, illumination, etc. Similarly, it could be possible to detect possible misbehaviours or failures using diagnostic tools in order to anticipate when human intervention is needed. Another field of applications where this could have applicability would be the e-health services or the support to the elderly and handicapped people in domestic tasks. However, the application scenario in which the Internet of Things concept can give the most important gains in terms of cost reduction and optimization is related logistics (logistics in the production process or in the transportation system). In fact, the possibility of keeping under control the position of each single item can really make the difference in the management decisions and, indeed, most of industrial interest into the Internet of Things is focusing on the applications in the logistic fields.

On the other hand, it is obvious that such a pervasive computing and communication infrastructure envisioned by the Internet of Things raises several concerns regarding privacy. In fact, a smart environment can collect a large amount of personal information that can be used to support added-value context aware services, but also to control people. The lots of talking and complaining about the announcement by an Italian retailer to tag their clothes with lifelasting RFID, demonstrates that people are not available to accept such technologies when they become a serious menace for their privacy. Accordingly, the Internet of Things will include several functionalities protecting personal information and ensuring that it will be used only for the purposes authorized by the information owner.

Finally, we can gain a completely different perspective if we look at the Internet of Things along with information-centric networking, which is expected to be one of the other major components of the Future Internet. The key rationale behind information centric networking is that users look at the Internet for the information it provides and are not interested in the specific host that stores this information. Unfortunately, IP-based Internet has been developed around the host-to-host communication paradigm, which does not fit this novel idea of the Internet.

It is useful to analyze how the Internet of Things can coexist and even exploit information-centric networking in the Future Internet. The first attempt in this direction is based on the concept that each resource is described by a metadata called Versatile Digital Item (VDI), decoupled by the resource and usable to represent *real world objects*. VDIs for real world objects contain information about the current state of the thing, for example, current owner, current usage type, position, etc. This approach has two important advantages:

- position of VDIs does not necessarily change as things move. This simplifies mobility management significantly;
- VDIs can implement functionalities which, accordingly, will run in infrastructure nodes instead of the thing objects they describe. This is important as processing capabilities available in actual things are extremely limited.

The major issues related to such an approach regard the naming scheme to be utilized and the related routing protocol. In general, in information centric networking contexts we can distinguish two approaches for what concerns the naming strategy: *flat* and *hierarchical* naming.

Usually, flat naming schemes are utilized along with *rendezvous node* based schemes, in which the name is mapped to a node where the state information of the *thing* is stored through a well known hash function. This rendezvous node will receive all queries and all updates regarding the VDI. Routers will route the query to the rendezvous node by appropriate processing of such a name. However, flat names are not mnemonic and therefore, appropriate services/applications are needed to *remember* such names, which introduces further complexity in the system. Furthermore, it is difficult (or even impossible) to control the selection of the rendezvous node.

A different approach that exploits hierarchical naming in information centric networks has been proposed in [44]. In this case, routers maintain updated routing tables in which the entry key is not the IP address but the name of the resource. Using this approach, the routing tables increase with the number of resources introduced in the network. In a global Internet of Things, this would be a manifest weakness. However, observe that the weakness may be overcome by using a structured naming that allows keeping the routing table size as small as possible.

6 Extending networking concepts to nanoscale communications

When trying to extrapolate networking problems to the future decades, with a long term horizon in mind, one of the emerging fields that is gaining momentum in the last years is the evolution of nano-technologies towards the "nanonetwork" concept, driven by the inspiration of applying biological systems and processes to networking technologies as it has been used in other fields such as optimisation, evolutionary and adaptive systems (e.g. genetic algorithms, ant colonies, etc.).

The term "nanotechnology" was first defined in [45] as the processing of, separation, consolidation, and deformation of materials by one atom or by one molecule, and this basic idea was further explored in more depth in the 1980s, with an acceleration of the activity in the field in the early 2000s. Nanotechnology enables the miniaturization and fabrication of devices in a scale ranging from 1 to 100 nanometers, considering a nano-machine as the most basic functional unit. Nano-machines are tiny components consisting of an arranged set of molecules which are able to perform very simple computation, sensing and/or actuation tasks [46]. Nano-machines can be further used as building blocks for the development of more complex systems such as nano-robots and computing devices such as nano-processors, nano-memory or nano-clocks.

In [47], the term "nanonetworks" is used to refer to the interconnection of nano-machines based on molecular communication. In general terms, a nano-machine is defined as "a device, consisting of nano-scale components, able to perform a specific task at nano-level, such as communicating, computing, data storing, sensing and/or actuation". The tasks performed by one nano-machine are very simple and restricted to its close environment due to its low complexity and small size. Then, the formation of a nanonetwork allows the execution of collaborative tasks, thus expanding the capabilities and applications of single nano-machines. For example, in case of nano-machines such as chemical sensors, nano-valves, nano-switches, or molecular elevators, the exchange of information and commands between them will allow them to work in a cooperative and synchronous manner to perform more complex tasks such as in-body drug delivery or disease treatments. Similarly, nanonetworks will allow extending the limited workspace of a single nano-machine enabling the interaction with remote nano-machines by means of broadcasting and multihop communication mechanisms.

Nanonetworks offer a wide range of application domains from nano-robotics to future health-care systems enabling the possibility to build from nano-scale body area networks to nano-scale molecular computers. Potential applications of nanonetworks are classified in [47] in four groups: biomedical, environmental, industrial and military applications, although they could also be used in other fields such as consumer electronics, since nanotechnologies have a key role in the manufacturing process of several devices.

Due to the small size of nano-machines that may be composed of just several moles of atoms or molecules in the orders of a few nanometers, traditional wireless communication with radio waves cannot be used for the communication. As a result, communication in nanonetworks is mainly to be realized through molecular communication means, making use of molecules as messages between transmitters and receivers. Two different and complementary coding techniques are explained in [47] to represent the information in nanonetworks. The first one, similar to what is done by traditional networks, uses temporal sequences to encode the information, such as the temporal concentration of specific molecules in the medium. According to the level of the concentration, i.e., the number of molecules per volume, the receiver decodes the received message, in a similar way as how the Central Nervous System propagates the neural impulses. The second technique is called molecular encoding and uses internal parameters of the molecules to encode the information such as the chemical structure, relative positioning of molecular elements or polarization. In this case, the receiver must be able to detect these specific molecules to decode the information. This technique is similar to the use of encrypted packets in communication networks, in which only the intended receiver is capable to read the information. In nature, molecular encoding is used in pheromonal communication, where only members of the transmitter specie can decode the transmitted message.

As noted above, nanonetworks cannot be considered a simple extension of traditional communication networks at the nano-scale, but they are a completely new communication paradigm, in which most of the communication processes are inspired by biological systems found in nature. A part from the differences in how the information is encoded, also the communication medium and the channel characteristics are different from those used in classical networks. Propagation speed differs from traditional communication networks, given that the information contained in molecules has to be physically transported from the transmitter to the receiver, and also propagation can be affected by random diffusion processes and environmental conditions, resulting in a much lower propagation speed. With respect to the transmitted information, as a difference from traditional communication networks, where information typically corresponds to text, voice or video, in nanonetworks, being the message a molecule, information is more related to phenomena, chemical states and processes. Correspondingly, most of the processes associated to nanonetworks are chemically driven, resulting in low power consumption.

As a result of the above considerations, most of the knowledge from existing communication networks is not suitable for nanonetworks, thus requiring innovative networking solutions according to the characteristics of the network components and the molecular communication processes. While some research efforts have been already carried out, many open research issues remain to be addressed for a practical realization of nanonetworks [48]. Firstly, further exploration of biological systems, communications and processes, should allow the identification of efficient and practical communication techniques to be exploited. Furthermore, it is to be studied the possible applicability of the definitions, performance metrics and basic techniques of classical networks, such as multiplexing, access, routing, error control, congestion, etc. With respect to the molecular communication, there are many open questions regarding the transmission and reception such as how to acquire new molecules and modify them to encode the information, how to manage the received molecules, and how to control the binding processes to release the information molecules to the medium. Similarly, communication reliability is another key issue in nanonetworks, since some applications will require reliable mechanisms to monitor or interrupt selective transmissions or on-going processes, while at the same time some molecular communication schemes are subject to random processes preventing to guarantee correct reception of transmitted messages. Consequently, communication reliability should be investigated by considering random communication features and particular infrastructures in molecular communication. Once the basic nanonetwork components are built, the transmission is controlled and the propagation is understood, advanced networking knowledge can be applied to design and realize more complex nanonetworks, even using layered architectures including medium access protocols, routing schemes and application interfaces [47].

As stated in [48], different studies have been performed on the design of nanoscale communication, introducing design approaches for molecular communication [46,49,50]. Other studies address the single molecular communication channel from an information theoretical perspective [51] and an adaptive error compensation mechanism for improving molecular communication channel capacity [52]. Molecular multiple-access, relay and broadcast channels are modelled and analyzed in terms of capacity limits in [53]. Although the first theoretical studies on nanonetworks are already there, there is still a lot of work to do before this promising field can become a reality. An interdisciplinary scientific approach involving networking concepts and biological aspects is needed to address the identified research challenges.

7 Conclusions

This section addresses heterogeneous and opportunistic wireless networks, providing some views on the possible evolution of future wireless networks, on some of the envisaged elements that will guide this evolution, and on the design challenges that are currently at stake.

The implications of network heterogeneity on how wireless networks will be operated and perceived by the users are presented, with the corresponding requirements for smart radio resource management strategies. The basic ingredients allowing to successfully coping with the increasing demand of traffic and bit rate are addressed: increase of the spectral efficiency, i.e., the number of bits/s that can be delivered per unit of spectrum; increase of the number of base stations to provide a service in a given area; and increasing the total available bandwidth. Within this area, the heterogeneity in networks and devices is also addressed, together with the multiplicity of data services with high bit rate requirements. Trends towards decentralisation and flat architectures, and perspectives on flexible spectrum management are shown as well.

Self-organising networks are included in the analysis, since it is expected that many mechanisms will be established to allow the self-configuration of a network with minimum human intervention. The features to be included in such an approach include: self-configuration, self-planning, self-optimisation, self-managing, and self-healing.

Opportunistic networks are also addressed, linking with delay-tolerant networks (among others), vehicular communication networks are analysed, linking with Intelligent Transportation Systems, and the Future Internet is viewed from the Internet of Things and RFID perspectives. Finally, a long term perspective is given towards the futuristic concept of the nanonetworks that target the extension of networking concepts towards the nano-scale communications.

References

1. ABI Research. Femtocell access points: Fixed-mobile convergence for residential, smb, and enterprise markets. Available: http://www.abiresearch.com (2006).
2. Tölli A., Hakalin P., Holma H.: Performance evaluation of common radio resource management (crrm). In: Proc. IEEE Int. Conf. Commun., New York, NY, pp. 3429–3433 (2002).
3. Pérez-Romero J., Sallent O., Agustí R.: On the optimum traffic allocation in heterogeneous CDMA/TDMA networks. IEEE Trans. Wireless Commun. **6**(9), pp. 3170–3174 (2007).
4. Pérez-Romero J., Sallent O., Agustí R., Díaz-Guerra M.A.: Radio resource management strategies in UMTS. Wiley & Sons, New York, NY (2005).
5. Zander J.: Radio resource management in future wireless networks: requirements and limitations. IEEE Commun. Mag. **35**(8), pp. 30–36 (1997).
6. Sallent O., Pérez-Romero J., Agustí R., Casadevall F.: Provisioning multimedia wireless networks for better QoS: RRM strategies for 3G WCDMA. IEEE Commun. Mag. **41**(2), pp. 100–106 (2003).
7. Hoffmeyer J.A.: Regulatory and standardization aspects of dsa technologies – global requirements and perspectives. In: Proc. IEEE Int. Symp. New Frontiers in Dynamic Spectrum Access Networks, Baltimore, MD (2005).
8. Buddhikot M., Ryan K.: Spectrum management in coordinated dynamic spectrum access based cellular networks. In: Proc. IEEE Int. Symp. New Frontiers in Dynamic Spectrum Access Networks, Baltimore, MD (2005).
9. Xiao F.H.Y (Ed.): Cognitive Radio Networks. CRC Press, Boca Raton, FL (2009).
10. Mueck M., et al.: ETSI reconfigurable radio systems: Status and future directions on software defined radio and cognitive radio standards. IEEE Commun. Mag. **48**(9), pp. 78–86 (2010).
11. UMTS Forum: Annual Report 2008 and Directions for 2009 (2009).
12. 3GPP Telecommunication Management: Self-Organizing Networks (SON), Concepts and Requirements (Release 8). Tech. Rep. (2008).
13. Next Generation Mobile Networks (NGMN): NGMN Recommendation on SON and O&M Requirements (2008).
14. Bogenfeld E., Gaspard I. (Eds.): Self-x in Radio Access Networks (2008).
15. Next Generation Mobile Networks (NGMN): NGMN Use Cases related to Self Organising Network, Overall Description (2007).
16. Schmelz L., der Berg J.V., Litjens R., Zetterberg K., Amirijoo M., Spaey K., Balan I., Scully N., Stefanski S.: Self-organisation in wireless networks – use cases and their interrelation. In: Proc. WWRF Meeting, Paris, France (2009).
17. 3GPP Telecommunication Management: Evolved Universal Terrestrial Radio Access Network (E-UTRAN). Self-configuring and self-optimizing network use cases and solutions. Tech. Rep. (2010).
18. Amirijoo M., Jorguseski L., Kurner T., Litjens R., Neuland M., Schmelz L., Turke U.: Cell outage management in LTE networks.
19. Prehofer C., Bettstetter C.: Self-organization in communication networks: Principles and design paradigms. IEEE Commun. Mag. **43**(7), pp. 78–85 (2005).
20. Hu H., Zheng X., Yang Y.: Self-configuration and self-optimization for LTE networks. IEEE Commun. Mag. **48**(2), pp. 94–100 (2010).

21. Sallent O., Pérez-Romero J., Sánchez-González J., Agustí R., Díaz-Guerra M.A., Henche D., Paul D.: A roadmap from UMTS optimisation to LTE self-optimisation, to appear.
22. Mitola, J. III, Maguire G.: Cognitive radio: Making software radios more personal. IEEE Personal Commun. **6**(4), pp. 13–18 (1999).
23. Mahmoud Q. (Ed.): Cognitive Networks: Towards Self-Aware Networks. Wiley and Sons. New York, NY (2007).
24. Thomas R., DaSilva L., MacKenzie A.: Cognitive networks. In: Proc. IEEE Int. Symp. New Frontiers in Dynamic Spectrum Access Networks, Baltimore, MD (2005).
25. Pantazopoulos P., Stavrakakis I., Passarella A., Conti M.: Efficient social-aware content placement for opportunistic networks. In: Proc. Int. Conf. Wireless On-demand Network Systems and Services, Kranjska Gora, Slovenia (2010).
26. Karlsson G., Lenders V., May M.: Delay-tolerant broadcasting. IEEE Trans. Broadcasting **53**(1), pp. 369–381 (2007).
27. Karlsson G., Lenders V., May M., Wacha C.: Wireless ad hoc podcasting. In: Proc. IEEE Conf. Sensor, Mesh and Ad Hoc Communications and Networks (SECON), San Diego, CA (2007).
28. Cenceme website. [Online]. Available: http://cenceme.org/
29. Metrosense website. [Online]. Available: http://metrosense.cs.dartmouth.edu/
30. Crawdad website. [Online]. Available: http://crawdad.cs.dartmouth.edu/
31. World Health Organization, World report on road traffic injury prevention, 2004.
32. Vehicle infrastructure integration (vii). [Online]. Available: http://www.its.dot.gov/itsnews/fact_sheets/vii.htm
33. Dot's intellidrive program. [Online]. Available: http://www.intellidriveusa.org/whoweare/program.php
34. ETSI intelligent transport systems. [Online]. Available: http://www.etsi.org/WebSite/Technologies/IntelligentTransportSystems.aspx
35. ITS project web site. [Online]. Available: http://www.its.dot.gov/index.htm
36. ITS standards on the move – etsi. [Online]. Available: http://www.etsi.org/WebSite/document/Technologies/ETSIpresentations
37. IEEE 802.11 Wireless LAN Working Group, Wlan mac and phy specifications: Wireless access in vehicular environments (wave). Tech. Rep. (2005).
38. IEEE 1609 – family of standards for wireless access in vehicular environments (wave). [Online]. Available: http://www.standards.its.dot.gov/fact_sheet.asp?f=80
39. Gozalvez J., Sepulcre M.: Opportunistic technique for efficient wireless vehicular communications. IEEE Veh. Technol. Mag. **2**(4), pp. 33–39 (2007).
40. Gallagher B., Akatsuka H.: Wireless communications for vehicle safety: Radio link performance & wireless connectivity issues. IEEE Veh. Technol. Mag. **1**(4), pp. 4–24 (2006).
41. ITU-internet report 2005: Internet of things. [Online]. Available: http://www.itu.int/osg/spu/publications/internetofthings/
42. Yan L., Zhang J., Yang L., Ning H. (Eds.): The Internet of Things. From RF-ID to the Next Generation Pervasive Networked Systems. CRC Press, Boca Raton, FL (2008).
43. Atzori L., Iera A., Morabito G.: The internet of things: A survey. Computer Networks **54**(15), pp. 2787–2805 (2010).
44. Jacobson V., Smetters D.K., Thornton J.D., Plass M.F., Briggs N.H., Braynard R.L.: Networking named content. In: Proc. ACM Int. Conf. Emerging Networking Experiments and Technologies, Rome, Italy (2009).
45. Taniguchi N.: On the basic concept of nano-technology. In: Proc. Int. Conf. on Production Engineering, London, UK (1974).
46. Suda T., Moore M., Nakano T., Egashira R., Enomoto A.: Exploratory research on molecular communication between nanomachines. In: Proc. Genetic and Evolutionary Computation Conf., Washington, DC (2005).
47. Akyildiz I., Brunetti F., Blázquez C.: Nanonetworks: A new communication paradigm. Computer Networks **52**, pp. 2260–2279 (2008).
48. Dressler F., Akan O.: A survey on bio-inspired networking. Computer Networks **54**, pp. 881–900 (2010).

49. Moore M., Enomoto A., Nakano T., Egashira R., Suda T., Kayasuga A., Kojima H., Sakakibara H., Oiwa K.: A design of a molecular communication system for nanomachines using molecular motors. In: Proc. IEEE Int. Conf. Pervasive Computing and Communications, Pisa, Italy (2006).
50. Nakano T., Suda T., Moore M., Egashira R., Enomoto A., Arima K.: Molecular communication for nanomachines using intercellular calcium signaling. In: Proc. IEEE Conf. Nanotechnology, Nagoya, Japan (2005).
51. Atakan B., Akan O.: An information theoretical approach for molecular communication. In: Proc. IEEE/ACM Int. Conf. Bio-Inspired Models of Network, Information and Computing Systems, Budapest, Hungary (2007).
52. Atakan B., Akan O.B.: On channel capacity and error compensation in molecular communication. Springer Trans. Computational Systems Biology LNBI 5410, pp. 59–80 (2008).
53. Atakan B., Akan O.: On molecular multiple-access, broadcast, and relay channel in nanonetworks. In: Proc. IEEE/ACM Int. Conf. Bio-Inspired Models of Network, Information and Computing Systems, Hyogo, Japan (2008).

Cognitive and cooperative wireless networks

Sergio Palazzo (✉), Davide Dardari, Mischa Dohler, Sinan Gezici, Lorenza Giupponi, Marco Luise, Jordi Pérez Romero, Shlomo Shamai, Dominique Noguet, Christophe Moy, and Gerd Asheid

1 Cognitive radio networks

The traditional approach of dealing with spectrum management in wireless communications has been the definition of a licensed user granted with exclusive exploitation rights for a specific frequency. While it is relatively easy in this case to ensure that excessive interference does not occur, this approach is unlikely to achieve the objective to maximize the value of spectrum, and in fact recent spectrum measurements carried out worldwide have revealed a significant spectrum underutilization, in spite of the fact that spectrum scarcity is claimed when trying to find bands where new systems can be allocated. Just to mention some examples of measurements, different studies can be found in [1–6], revealing that overall occupation in some studies for frequencies up to 7 GHz could be in the order of only 18%.

As a result of the above, one of the current research trends in the spectrum management are the so-called Dynamic Spectrum Access Networks (DSANs), in which unlicensed radios, denoted in this context as Secondary Users (SUs) are allowed to operate in licensed bands provided that no harmful interference is caused to the licensees, denoted in this context as Primary Users (PU). The proposition of the TV band Notice of Proposed Rule Making (NPRM) [7], allowing this secondary operation in the TV broadcast bands if no interference is caused to TV receivers, was a first milestone in this direction. In this approach, SUs will require to properly detecting the existence of PU transmissions and should be able to adapt to the varying spectrum conditions, ensuring that the primary rights are preserved. These events culminated in the creation of the IEEE 802.22, developing a cognitive radio-based physical and medium access control layer for use by license-exempt devices on a non-interfering basis in spectrum that is allocated to the TV broadcast service. Based on these developments it is reasonable to think that the trend towards DSANs has just started and that given the requirements for a more efficient spectrum usage, it can become one of the important revolutions in the wireless networks for the next decades, since it completely breaks the way how spectrum has been traditionally managed.

Benedetto S., Correia L.M., Luise M. (eds.): The Newcom++ Vision Book. Perspectives of Research on Wireless Communications in Europe.
DOI 10.1007/978-88-470-1983-6_4,

The primary-secondary (P-S) spectrum sharing can take the form of cooperation or coexistence. Cooperation involves explicit communication and coordination between primary and secondary systems, and coexistence means there is none. When sharing is based on coexistence, secondary devices are essentially invisible to the primary. Thus, all of the complexity of sharing is born by the secondary and no changes to the primary system are needed. There can be different forms of coexistence, such as spectrum underlay (e.g. UWB) or spectrum overlay (e.g. opportunistic exploitation of white spaces in spatial-temporal domain sustained on spectrum sensing, coordination with peers and fast spectrum handover). As for cooperation, again different forms of P-S interactions are possible. For example, spatial-temporal white spaces that can be exploited by SUs can be signalled through appropriate channels. In addition, the interaction between PUs and SUs provides an opportunity for the license-holder to demand payment according to the different quality of service grades offered to SUs.

One of the key enabling technologies for DSAN development is Cognitive Radio (CR), which has been claimed to be an adequate solution to the existing conflicts between spectrum demand growth and spectrum underutilization. The term Cognitive Radio was originally coined by J. Mitola III in [8] and envisaged a radio able to sense and be aware of its operational environment so that it can dynamically and autonomously adjust its radio operating parameters accordingly to adapt to the different situations. CR concept was built in turn upon the Software Defined Radio (SDR) concept, which can be understood as a multiband radio supporting multiple air interfaces and protocols and being reconfigurable through software run on DSP or general-purpose microprocessors. Consequently, SDR constituted the basis for the physical implementation of CR concepts.

Thanks to this capability of being aware of actual transmissions across a wide bandwidth and to adapt their own transmissions to the characteristics of the spectrum, CRs offer great potential for bringing DSANs to reality, and in fact DSANs are usually referred to as Cognitive Radio Networks (CRN). The operating principle of a CR in the context of a DSAN is to identify spatial and temporal spectrum gaps not occupied by primary/licensed users, usually referred to as *spectrum holes* or *white spaces*, place secondary/unlicensed transmissions during such spaces and vacate the channel as soon as primary users return. The CR concept therefore implicitly relies on two basic premises: the existence of enough white spaces caused by primary spectrum underutilization and the ability of secondary users to effectively detect and identify the presence of different licensed technologies in order not to cause harmful interference.

From a general operation perspective, a CR follows the so-called *cognition cycle* to enable the interaction with the environment and the corresponding adaptation. It consists in the *observation* of the environment, the *orientation* and *planning* that leads to making the appropriate *decisions* pursuing specific operation goals, and finally *acting* over the environment. Decisions on the other hand can be reinforced by *learning* procedures based on the analysis of prior observations and on the corresponding results of the prior actuations. Then, when particularizing the cognition cycle to the dynamic spectrum access for a secondary user, the observation turns out to be the spectrum sensing in order to identify the potential white spaces, the orientation and planning steps would be associated with the analysis of the available white spaces, and finally the acting step would be in charge of selecting the adequate white space to make the secondary transmission, together with the setting of the appropriate radio parameters such as transmit power, modulation formats, etc.

There are a number of techniques to be developed for an implementation of efficient secondary spectrum usage through cognitive radio networks, and are classified in [9] as spectrum sensing, spectrum management, spectrum mobility and spectrum sharing mechanisms. These techniques are briefly discussed in the following:

- *Spectrum sensing.* It consists in detecting the unused spectrum bands that can be potentially exploited for secondary communications. A lot of different spectrum sensing techniques have been studied in the last years, such as the energy detector, which does not include any specific knowledge about the primary signal to be detected, the matched filter detection, which requires the knowledge of the specific primary signal formats, or the cyclostationarity feature detection. Also the possibility of combining sensing measurements from different sensors through appropriate fusion schemes has been considered in the so-called cooperative sensing. Even from a more general perspective, the possibility that the network provides knowledge about the current spectrum bands available through some control channel has also been considered. This is the case of e.g. the development of the so-called Cognitive Pilot Channel (CPC) in [10]. From this perspective, and having in mind the possibility of combining the knowledge provided by the network with the knowledge acquired by the sensing process, spectrum sensing concept can be generalised to the concept of spectrum awareness.
- *Spectrum management.* This refers to the selection of the most adequate spectrum band to carry out the transmission in accordance with the secondary user requirements. This selection should be made based on the characteristics of the channel in terms of e.g. the maximum capacity that can be obtained by the secondary users, and also taking into consideration the maximum interference that can be tolerated by primary receivers. The decision making process here can be benefited from the application of learning strategies, that, based on experience acquired from prior decisions, can orient the decisions towards the selection of some channels in front of others. As an example, in case that in some channels the primary user activity is higher, it is more likely that primary users appear forcing the secondary transmitter to evacuate the channel, so if such knowledge was available, this could prevent the secondary network from selecting these channels.
- *Spectrum mobility.* This functionality consists in establishing appropriate mechanisms to ensure that an on-going secondary communication can be continued whenever a primary user appears in the occupied bandwidth. This will thus involve the ability to detect the appearance of this primary user, which requires some continuous monitoring of the channel, e.g. through sensing mechanisms. Then, when the primary user appears, the occupied channel has to be evacuated, and an alternative spectrum piece has to be found where the communication can be reassumed, which is usually called *spectrum handover.*

- *Spectrum sharing.* This function targets the provision an efficient mechanism so that coexisting secondary users can share the available spectrum holes. Adequate Medium Access Control (MAC) protocols and scheduling mechanisms would be needed, and they would be very much dependant on how the secondary network is deployed, e.g. if it is infrastructure or infrastructure-less based, etc.

Although all the above functions have become a hot research topic during the last few years, there is still a lot of work to do before CRNs can become a reality. This will involve not only technical aspects, but also significant regulatory changes will be needed. In addition, this will also have implications from the techno-economic perspective, with the appearance of new business models to exploit the capabilities offered by CRNs, involving different possibilities ranging from secondary cellular operators that could offer services at cheaper prices at the expense of a somehow reduced quality, to the deployment infrastructure-less secondary networks that would enable the communication of short range devices. Clearly, all these elements put a quite long-term perspective, maybe of several decades, before the final implementation of CRNs.

2 Cognitive positioning

2.1 Introduction

According to the definition above, cognitive systems strive for optimum spectrum efficiency by allocating capacity as requested in different, possibly disjoint frequency bands. Such approach is naturally enabled, by the adoption of flexible MultiCarrier (MC) technologies, in all of its flavors: traditional OFDM, Filter-Bank Multicarrier Modulation (FBMCM) [20], and possibly non-orthogonal formats with full time/frequency resource allocation. Most current and forthcoming wideband standards for wireless communications (3GPP's Long-Term-Evolution is a paradigmatic example in this respect) are based on such multicarrier signalling technology, so that the signal allocated to each terminal is formed as the collection of multiple data symbols intentionally scattered across non-contiguous spectral chunks.

On the other hand, modern wireless networks more and more expect availability of location information about the wireless terminals, driven by requirements coming from applications, or just for better network resources allocation. Thus, signal-intrinsic capability for accurate localization is a goal of 4th Generation (4G) as well as Beyond-4G (B4G) networks. All signal processing techniques that can contribute to the provision of accurate location information are welcome in this respect. Such techniques can pair the ones that a cognitive terminal adopts to establish a reliable, high-capacity link.

The most accurate techniques to perform localization of a wireless terminal are based on time-of-arrival (TOA) estimation of a few radio ranging signals, followed by ranging and appropriate triangulation. Therefore, the precision of positioning is strictly related to the accuracy that can be attained in the estimation of the propagation delay of the radio signal [16, 18]. In the following, we will see that a multicarrier signal format, possibly split in (two or more) non-contiguous bands, gives new opportunities in terms of enhanced-accuracy time delay estimation (that ultimately translates into enhanced accuracy positioning). By chance, the two issues of super-accurate signal TOA estimation and sparse multicarrier resource allocation, we can say, "marry in heaven".

To let the reader understand how this could be done, we will start with a review of the Modified Cramér-Rao bound (MCRB), its frequency-domain computation, as well as the study of the impact on the bound of the location of the received signal spectrum within the receiver bandwidth. After this, we will discuss how to optimize an MC signal format through minimization of the MCRB, to come to the description of Cognitive Positioning (CP) opportunities [12,21].

2.2 *Criteria to optimize the function of ranging*

The Cramér-Rao bound (CRB) is a fundamental lower bound on the variance of any estimator [13, 19] and, as such, it serves as a benchmark for the performance of actual parameter estimators [11, 15, 17]. It is well known and widely adopted for its simple computation, but its close-form evaluation becomes mathematically intractable when the vector of observables contains, in addition to the parameter to be estimated, also some *nuisance parameters*, i.e., other unknown random quantities whose values are not the subject of the estimator (information data, random chips of the code of a ranging signal, etc.), but that concur to shape the actual observed waveform. The MCRB for estimation of the TOA (delay) τ for a received signal $x(t-\tau)$ embedded in complex-valued AWGN $n(t)$ with two-sided psd $2N_0$ is found to be [14]

$$\mathrm{MCRB}(\tau) = \frac{N_0}{\mathbb{E}_{\mathbf{c}}\left\{\int_{T_{\mathrm{obs}}}\left|\frac{\partial x(t-\tau)}{\partial \tau}\right|^2 dt\right\}}, \tag{1}$$

where $\mathbf{c}$ is a vector collecting all of the nuisance parameters, T_{obs} is the observation time-interval, and $\mathbb{E}_{\mathbf{c}}$ indicates statistical expectation wrt $\mathbf{c}$. This expression holds in the case of ideal coherent demodulation, i.e., assuming that during signal tracking the carrier frequency and the carrier phase are known to a sufficient accuracy. From this expression, we can devise a simple criterion for optimal signal design: finding that specific waveform that, across a pre-set bandwidth and for a certain signal-to-noise-ratio (SNR), gives the minimum MCRB value. We will not consider here any aspects related to a possible *bias* of the estimator arising in a severe multipath propagation environment, to concentrate on the main issue that we have just stated.

Assume we are receiving a generic pilot ranging signal $x(t;\mathbf{c})$ bearing no information data, but containing a pseudo-random ranging code $\mathbf{c}$, whose chips $\in \{\pm 1\}$ are considered as binary iid nuisance parameters. This signal turns out to be a *parametric random process*, for which each time-unlimited sample function is a signal with finite power P_x and chip rate $R_c = 1/T_c$.

Assuming that the observation time is sufficiently large, the MCRB can be also derived through frequency-domain computation [21]:

$$\mathrm{MCRB}(\tau) = \frac{N_0}{T_{\mathrm{obs}} 4\pi^2 \int_{-\infty}^{+\infty} f^2 \mathscr{P}_x(f)\, df} = \frac{1}{8\pi^2 N \beta^2 \cdot E_c/N_0}, \tag{2}$$

where $\mathscr{P}_x(f)$ is the power spectral density (PSD) of $x(t)$, E_c is the received signal energy per chip, $NT_c = T_{\mathrm{obs}}$, and β^2 is the normalized second-order moment of the PSD,

$$\beta^2 = \frac{1}{2P_x} \int_{-\infty}^{+\infty} f^2 \mathscr{P}_x(f)\, df. \tag{3}$$

We conclude that the MCRB depends on the second-order moment β^2 of the PSD of the signal x, irrespective of the type of signal format (modulation, spreading, etc.) that is adopted.

2.3 Ranging multicarrier signals

A ranging MC signal can be constructed following the general arrangement of multicarrier (MC) modulation: the input chip stream $c[i]$ of the ranging code is parallelized into N substreams with an MC symbol rate $R_s = R_c/N = 1/(NT_c) = 1/T_s$, where T_s is the time duration of the "slow-motion" ranging chips in the N parallel substreams. We can use a "polyphase" notation for the k-th ranging subcode ($k = 0, 1, \ldots, N-1$) in the k-th substream (subcarrier) as $c^{(k)}[n] \triangleq c[nN + k]$, where k is the subcode identifier and/or the subcarrier index, whilst n is a time index that addresses the n-th MC symbol (block) of time length $T_s = NT_c$. The substreams are then modulated onto a raster of evenly-spaced subcarriers with frequency spacing f_{sc} and the resulting modulated signals are added to give the (baseband equivalent of the) overall ranging signal. In Filter-Bank Multicarrier Modulation (FBMCM) (also called Filtered MultiTone, FMT) the spectra on each subcarrier are strictly bandlimited and nonoverlapping, akin to conventional single channel per carrier (SCPC). The resulting signal is [20]

$$x(t) = \sqrt{\frac{2P_x}{N}} \sum_n \sum_{k=0}^{N-1} p_k c^{(k)}[n] g(t - nT_s) \exp[j2\pi k f_{\mathrm{sc}} t], \tag{4}$$

where $g(t)$ is a bandlimited pulse, for instance a square-root raised cosine pulse with roll-off factor α. In this case, the subcarrier spacing is $f_{\mathrm{sc}} = (1+\alpha)/T_s$. It is well known that this arrangement has an efficient realization based on IFFT processing followed by a suited polyphase filterbank based on the prototype filter $g(t)$. The real-valued coefficient p_k in (4), $0 \le p_k \le N$, $\sum_{k=0}^{N-1} p_k^2 = N$, allows us to perform the function of *power allocation* by allowing different amounts of signal power to be placed on the different subcarriers. Some p_k's can also be 0 indicating that the relevant subcarriers (or even a whole subband) is not being used. We will see how this feature is essential for the characteristics of cognitive positioning.

When T_{obs} is sufficiently large, $T_{\mathrm{obs}} = N_m T_s$, $N_m \gg 1$, the MCRB for the MC signal can be easily computed:

$$\mathrm{MCRB}(\tau) = \frac{T_c^2}{8\pi^2 \cdot \frac{E_c}{N_0} \cdot \frac{N_m}{N} \cdot \left[\xi_g + \frac{(1+\alpha)^2}{N} \sum_k p_k^2 k^2\right]}, \tag{5}$$

where ξ_g is the so-called pulse shape factor (PSF),[1] a normalized version of the Gabor bandwidth of pulse $g(t)$:

$$\xi_g \triangleq \frac{T_s^2 \int_{-\infty}^{+\infty} f^2 |G(f)|^2 df}{\int_{-\infty}^{+\infty} f^2 |G(f)|^2 df}. \tag{6}$$

2.4 Cognitive signals for cognitive positioning

An MC signal with uneven, adaptive power distribution can be adopted to implement a *cognitive positioning system*. In our envisioned FBMCM scheme for positioning, the proper power allocation allows to reach the desired positioning accuracy, not only in an additive white Gaussian noise (AWGN) channel, but also in an additive coloured Gaussian noise (ACGN) channel. Coloured noise arises from variable levels of interference produced by co-existing (possibly primary) systems on different frequency bands. The key assumption is that such interference can be modelled as a Gaussian process. This is certainly justified in wireless networks with unregulated multiple access techniques such as CDMA and/or UWB.

To be specific, let us investigate the issue of finding the power allocation scheme that gives the *minimum MCRB* for TDE in a Gaussian channel whose (additive) noise has a variable PSD $\mathscr{P}_n(f)$. After some algebra, we find out that

$$\mathrm{MCRB}(\tau)|_{\mathrm{ACGN}} = \frac{1}{4\pi^2 T_{\mathrm{obs}} (\Delta f)^3} \left(\sum_k k^2 \frac{\mathscr{P}_x(k\Delta f)}{\mathscr{P}_n(k\Delta f)} \right)^{-1}, \tag{7}$$

where $\Delta f = f_{\mathrm{sc}}$ is the subcarrier spacing, and where the PSD of the transmitted signal and of the noise were considered constant across each subband. A fundamental result is obtained if we let $N \to \infty$ (thus $\Delta f \to df$):

$$\mathrm{MCRB}(\tau)|_{\mathrm{ACGN}} = \frac{1}{4\pi^2 T_{\mathrm{obs}}} \left(\int_{-B}^{+B} f^2 \frac{\mathscr{P}_x(f)}{\mathscr{P}_n(f)} df \right)^{-1}, \tag{8}$$

where B is the (finite) signal bandwidth.

Coming back to the problem of enhancing TDE accuracy, and sticking for simplicity to the finite-subcarriers version of the problem, we have to minimize the MCRB (7) with a constraint on the total signal power P_x.

[1] For a square-root raised-cosine pulse with roll-off factor α, $\xi_g = 1/12 + \alpha^2 (1/4 - 2/\pi^2)$.

Considering that $\mathscr{P}_x(k\Delta f)$ is proportional to P_k^2, we are to find the power distribution P_k^2 that *maximizes*

$$\sum_k p_k^2 \frac{k^2}{\mathscr{P}_n(k\Delta f)} \tag{9}$$

subject to the constraints $\sum_k p_k^2 = N$ and, of course, $p_k \geq 0$. The optimal distribution is easily found to be

$$\begin{cases} p_k = 0, & k \neq k_M, \\ p_k = \sqrt{N}, & k = k_M \end{cases} \qquad k_M = \arg\max_k \frac{k^2}{\mathscr{P}_n(k\Delta f)} \tag{10}$$

that corresponds to placing all the power onto the sub-band for which the *squared-frequency to noise ratio* (SFNR) $k^2 / \mathscr{P}_n(k\Delta f)$ is maximum. In AWGN, this is the band-edge subcarrier, as is known form Gabor bandwidth analysis.

A more realistic case study for CP in ACGN takes also into account possible power limitations on each subcarrier that prevents from concentrating all of the signal power onto the edge subcarriers (for AWGN) or on the subcarrier with the best SFNR as above. We add thus the further constraint $0 \leq p_k^2 \leq P_{\max} < N$. The solution to this new power allocation problem can be easily found via linear programming:

- order the square-frequency-to-noise-ratios SFNR_k from the highest to the lowest; set the currently allocated power to zero; mark all carriers available;
- find the available power as the difference between the total power N and the currently allocated power. If it is null, then STOP, else, if it is larger than $P_{\max}$, then put the maximum power $P_{\max}$ on the available carrier with the highest SFNR; else put on the same carrier the (residual) available power;
- update the currently allocated power by adding the one just allocated, and remove the just allocated carrier from the list of available carriers. If the list is empty, then STOP, else goto the previous step).

This results in a set of bounded-power subcarriers that gives the optimum power allocation (minimum TDE MCRB) with ACGN. An avenue for research is finding practical algorithms that attain the bounds above in a realistic environment, and extend the results above to cases with strong multipath.

3 Cooperative wireless networks

The concept of cooperation actually emerged in the late sixties with the work of Van Der Meulen about "Three terminal communications". Interestingly, the capacity of this scheme is still an open problem today. More recently, the concept of relaying or cooperation has gained a lot of interest for several reasons:

- In mobile or wireless communications, the potential offered by multi-antenna transmission and/or reception is now clearly established, and this technology, known under the generic name of MIMO, has found its way in a number of standards. "Classical" MIMO however assumes that the different antennas can be or are colocated on the same site. There are scenarios, however, where this assumption cannot be met for operational reasons and/or cost reasons. Therefore, the idea has emerged as whether different non colocated entities could *form a coalition* to mimic in a distributed manner a multi-antenna system, thereby getting access to the benefits of MIMO in term of rate (multiplexing gain) and/or diversity, according to the well-known trade-off between the two.

- A natural way to exploit this idea is to serve a user by means of two or more base stations. This concept is also known as macro-diversity. Assuming a wired backhaul, the base stations know in the best case all the data and the channel state information of all the users in the cluster or "supercell" served by the coordinated base-stations. There is no issue of *decoding strategy* by the *relay* in such a case because the wired nature of the backhaul makes sure the data are available without any error. The design of linear precoders and decoders for such a scenario, possibly robust to imperfect channel knowledge, is its infancy. Another issue, to avoid very heavy signalling in the backhaul, is that of distributed solutions based on partial data and/or channel knowledge at the coordinating node, yet approaching the performance of a fully *informed* solution.
- While the motivation behind the previous concept is mainly to avoid intercell interference (the COMP approach in LTE), another motivation is associated with the issue of coverage and *badly* located users which might be out of (good) reach by any base station. An emerging concept is that of a "popping up base station" with wireless backhaul, that would help one or many poor users. In this case, because of the wireless nature of the backhaul, issues similar to that of the classical relay channel reappear, which are of course related to the fact that the base station first has to receive properly the data to be relayed. Hence the issue of decoding strategy (in the broad sense) has to be considered. Along these lines there should be an increasing interest for the design of nodes serving as relays and which would be equipped with MIMO capability: MIMO relay schemes.
- Moving to a totally different scenario like wireless sensor networks, there is also a clear interest for cooperative solutions. As a matter of fact, sensing nodes may usually be equipped with a battery which is supposed to last as long as possible, may not be easily reachable and limits the communication capability of the node in term of available power. Therefore not all nodes are necessarily able to directly establish a connexion with the *fusion center* or the *collecting point*. The network is then of the mesh types rather than of the star type. The data has to reach the collecting node by means of multiple hops where some sensing nodes basically relay the information of their neighbours. The choice of the cooperating nodes may be based on several criteria or utility functions, incorporating not only rate and/or bit/packet error measures but also penalty depending on the power used and/or the status of the battery of the possibly cooperating nodes.
- These days there is an increasing interest for *cognitive communications systems*. The basic idea behind is the capability to sense the spectrum, to detect possible holes and to establish communication in *free frequency slots*. While the concept of *cognition* has mainly been discussed for the PHY layers, there is an emergence

of the concept of cognitive networks, where the load or presence of *available resources* would even be exploited at different layers of the systems. Coming back to the PHY layer, cooperation is a natural tool and a desirable feature in order to be able to properly sense the spectrum and address concerns like the hidden terminal problem. Therefore, obtaining *spectrum maps* naturally leads to a joint cooperative-cognitive approach.

Wireless systems are often networks in the sense that multiple entities or users compete for the available resource(s). For instance, a node in a multihop setup may be expected to relay the signals of several adjacent nodes. In that case, the strategy to be chosen by the relaying node to serve the neighbouring nodes has to be properly addressed. An important concept that emerged recently and deserves further investigation is that of *network coding* which shows promises but also needs to properly encode and simultaneously relay the information of several users at the same time.

A final remark has to do with energy saving and *green communications*. For a prescribed performance metric to be achieved at the receiving end, the combination of transmitter and relay might require a lower total transmission power than if the transmitter alone is sending this information. There are results clearly indicating this. However it should also be noted that transmission power is only one part of the global picture. Associated with any communicating nodes, there are additional power prices like those associated with computation, security, etc. It would be highly interesting and motivating to investigate the potential of relaying or cooperative communications at the light of a holistic analysis of power consumption.

4 Docitive radios & networks

4.1 Introduction

As already evidenced throughout this paper, cognitive radios and networks are perceived as a facilitator for improved efficiency of scarce spectrum access and management. Cognition, from *cognoscere = to know* in Latin, is typically defined as "a process involved in gaining knowledge and comprehension, including thinking, knowing, remembering, judging, and problem solving" [22]. Cognition has been the focus of numerous disciplines in the past, such as biology, biomedicine, telecommunications, computer science, etc. Across all these domains, emphasis has clearly been on a certain degree of intelligence which allows a cognitive system "to work properly under conditions it was initially not designed for" [23].

Said intelligence is typically accomplished by profoundly sensing the surroundings of the cognitive node and learning upon the acquired information [8]. This learning process is often a lengthy and complex process in itself, with complexity increasing with an increasing observation space. It is however needed to truly realize a cognitive radio as otherwise only opportunistic access is guaranteed at best. And whilst cognition and learning have received a considerable attention from various communities in the past, the process of knowledge transfer, i.e. teaching, however has received fairly little attention to date. This contribution thus aims at introducing a novel framework referred to as docitive radios, from *docere = to teach* in Latin, which relates to radios (or general entities) which teach other radios. These radios are not (only) supposed to teach them the end-result (e.g. in form of "the spectrum is occupied") but rather elements of the methods to getting there. This concept mimics well our society-driven pupil-teacher paradigm and is expected to yield significant benefits for cognitive and thus more efficient network operation. Important and unprecedented questions arise in this context, such as which information ought to be taught, what is the optimum ratio between docitive and cognitive radios, etc.

An illustrative example, which will be corroborated in more depth in a subsequent section, models a cognitive radio system as a multi-agent system, where the radios learn through the paradigm of multi-agent learning. A typical learning mechanism for single agent systems is Q-Learning, belonging to the class of Reinforcement Learning. When it comes to multi-agent systems, Q-learning can be adapted to this setting, by implementing decentralized Q-learning. In this case, each node has to build a state-action space where it needs to learn the optimal policy for taking actions in each state. Depending on the dimension of the state-action space, the training process may be extremely time consuming and complex. However, if nodes are instructed to learn some disjoint or random parts of the state-action space, then they can share the acquired knowledge with their neighboring nodes. This facilitates learning but does not yield the end-result per se.

4.2 *Brief taxonomy*

A high-level operational cycle of docitive radios is depicted in Fig. 1. It essentially extends the typical cognitive radio cycle [8] through the docitive teaching element, where each of these elements typically pertains to the following high-level issues:

- *Acquisition*. The acquisition of data is quintessential in obtaining sufficient information of the surrounding environment. This data can be obtained by means of numerous methods, such as sensing performed by the node itself and/or in

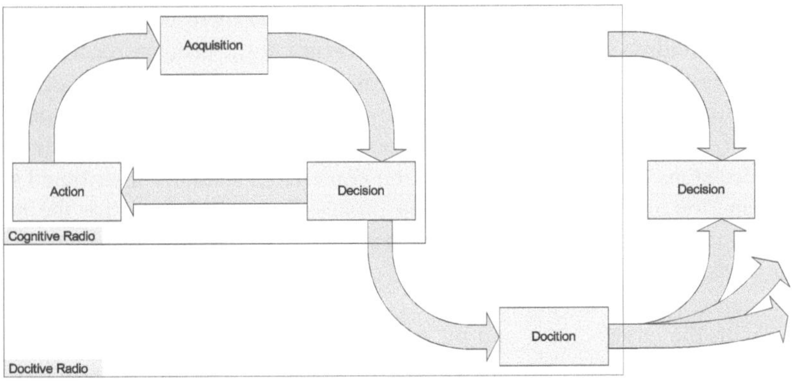

Fig. 1. Docitive cycle which extends the cognitive cycle through the teaching element

conjunction with spatially adjacent cooperating nodes; docitive information from neighboring nodes; environmental/docitive information from databases; etc.

- *(Intelligent) Decision.* The core of a cognitive radio is without doubt the intelligent decision engine which learns and draws decisions based on the provided information from the acquisition unit. The majority of cognitive devices today run some simple opportunistic decision-making algorithms; however, some more sophisticated learning and decision algorithms in form of e.g. unsupervised, supervised or reinforcement learning are available too.
- *Action.* With the decision taken, an important aspect of the cognitive radio is to ensure that the intelligent decisions are actually carried out, which is typically handled by a suitably reconfigurable software defined radio (SDR), some policy enforcement protocols, among others.
- *Docition.* An extension of the cognitive networking part is realized by means of an entity which facilitates knowledge dissemination and propagation, where so far rather end results have been shared (e.g. through cooperative sensing). A significant and non-trivial extension to this docitive paradigm comprises dissemination of information which facilitates learning.

4.3 Docitive example: Wireless multi-agent systems

We subsequently exemplify the operation of a docitive radio by means of wireless multi-agent learning systems. To use prior notion from the machine learning community, we will use the concept of agents which are defined as a computational mechanism that exhibits a high degree of autonomy performing actions, based on information from the environment. As a result of that, a multi-agent system is a complex system where multiple agents interact with one another, where the actions of each agent have impact on the environment of the others and where each agent has only partial information of the overall system [24]. The cognitive radio scenario can be easily mapped onto a multi-agent system [25], since it consists of multiple intelligent and autonomous agents, i.e. the cognitive radios, with the following characteristics: 1) the aggregated interference on a primary receiver depends on the independent decisions made by the multiple agents; 2) there is no central entity in charge of providing a global control of interference at the primary receivers coming from the multiple cognitive radios, so that the system architecture is decentralized; 3) a solution based on a centralized agent would not be scalable with respect to the number of cognitive radios; 4) the data based on which the cognitive radio system makes decisions about resource allocation come from spatially distributed sources of information and the decision making process is asynchronous for the multiple agents; 5) the individual decisions of each agent have to be self-adaptive depending on the decisions made by the other agents and on the surrounding environment. The above mentioned self-adaptation has to be achieved progressively by directly interacting with the environment and by properly utilizing the past experience, which is obtained through real-time operations. As a result, the common objective of the multiple agents is to distributively learn an optimal policy to achieve a common objective.

In case of single agent, the environment can be modeled as Markov Decision Process (MDP) [26], which is a tuple $\langle S, A, T, R \rangle$, where S is a finite set of environment states, A is a set of actions, $T : S \times A \times S \rightarrow [0,1]$ is the Markovian transition function that describes the probability $p(s'|s,a)$ of ending up in state s' when performing action a in state s, and $R : S \times A \rightarrow \mathbb{R}$ is a reward function that returns the reward obtained after taking action a in state s. An agent's policy is defined as a mapping $\pi : S \rightarrow A$. The objective is to find the optimal policy π^* that maximizes the expected discounted future reward $U^*(S) = max_{\pi}\mathbb{E}\left[\sum_{t=0}^{\infty} \gamma^t R(s_t)|\pi, s_0 = s\right]$, for each state s, and where s_t indicates the state at time step t. The expectation operator averages over reward and stochastic transitions and $\gamma \in [0,1)$ is the discount factor. We can also represent this using Q-values, which store the expected discounted future reward for each state s and possible action a:

$$Q^*(s,a) = R(s,a) + \gamma \sum_{s'} p(s'|s,a) \max_{a'} Q\left(s', a'\right), \tag{11}$$

where a' is the action to take in state s'. The optimal policy for a state s is the action $\arg\max_a Q^*(s,a)$ that maximizes the expected future discounted reward. Reinforcement learning can be applied to estimate $Q^*(s,a)$; in particular, Q-learning is a widely used reinforcement learning method when the transition model is unavailable [27,28]. This method starts with an initial estimate $Q(s,a)$ for each state action pair. When an action a is taken in state s, reward $R(s,a)$ is received and the next state s' is observed, the corresponding Q-value is updated by:

$$Q(S,a) = Q(s,a) + \alpha \left[R(s,a) + \gamma \cdot \max_{a'} Q(s',a') - Q(s,a)\right], \tag{12}$$

where $\alpha \in (0,1)$ is an appropriate learning rate. Q-learning is known to converge to the optimal $Q^*(s,a)$.

In case of multi-agent systems, all knowledge is not available locally in a single agent, but relevant knowledge, such as training experience and background information, is distributed among the agents within the system. In this case, we talk about distributed reinforcement learning, or multi-agent learning. The problem is how to ensure that individual decisions of the agents result in jointly optimal decisions for the group and how to reliably propagate this information over the wireless channel to spatially adjacent nodes. In principle, it should be possible to treat a multi-agent system as a single agent with complete information about the other agents, and learn the optimal joint policy using single-agent reinforcement learning techniques. However, both the state and action space scale exponentially with the number of agents, rendering this approach infeasible for most problems. Alternatively, we can let each agent learn its policy independently of the other agents, but then the transition model depends on the policy of the other learning agents, which may result in oscillatory behavior. This introduces game-theoretic issues to the learning process, which are not yet fully understood [29].

Contributions in literature [30] suggest that the performances of a multi-agent system can be improved by using cooperation among learners in a variety of ways.

In fact, it can be assumed that each agent does not need to learn everything by its own discover, but can take advantage of the exchange of information and knowledge with other agents or with more expert agents, thus leading to a teaching paradigm. It is demonstrated in [30] that if cooperation is done intelligently, each agent can benefit from other agents' information. Depending on the degree of cooperation among agents, we propose to consider the following cases for future studies:

- *Independent learners.* The agents do not cooperate, ignore the actions and rewards of the other agents in the system and learn their strategies independently. The standard convergence proof for Q-learning, in case of single agent system, does not hold in this case, since the transition model depends on the unknown policy of the other learning agents. In particular, each agent's adaptation to the environment can change the environment itself in a way that makes the other agents' adaptations invalid. Despite that, this method has been applied successfully in multiple cases.
- *Cooperative learners sharing state information.* The agents follow the paradigm of independent learning, but can share instantaneous information about their state. It is expected that sharing state information is beneficial in case that it is relevant and sufficient for learning.
- *Cooperative learners share policies or episodes.* The agents follow the paradigm of independent learning, but can share information about sequences of state, action and reward and learned decision policies corresponding to specific states. These episodes can be exchanged either among peers, or with more expert peers, i.e. teachers. It is expected that such cooperative agents can speed up learning, measured by the average number of learning iterations, and reduce the time for exploration, even though the asymptotic convergence can be reached also by independent agents.
- *Cooperative learners performing joint tasks.* Agents can share all the information required to cooperatively carry out a certain task. In this case the learning process may be longer, since the state-action space is bigger, but oscillatory behaviors are expected to be reduced.
- *Team learners.* The multi-agent system can be regarded as a single agent in which each joint action is represented as a single action. The optimal Q-values for the joint actions can be learned using standard single-agent Q-learning. In order to apply this approach, a central controller should model the MDP and communicate to each agent its individual actions, or all agents should model the complete MDP separately and select their individual actions. In this case, no communication is needed between the agents but they all have to observe the joint action and all individual rewards. The problem of exploration can be solved by using the same random number generator and the same seed for all agents. Although this approach leads to the optimal solution, it is infeasible for problems with many agents, since the joint action space, which is exponential with the number of agents, becomes intractable.

From the above, we incur that the concept of joint learning has received attention in recent years in the machine learning and artificial intelligence community; however, its application to cognitive radios operating primarily over a wireless broadcast channel has not been addressed by any study yet and, coupled with the potential gains, essentially inspired the concept of docitive radios.

4.4 Vision and challenges

Docitive radios and networks emphasize on the teaching mechanisms and capabilities of cognitive networks, and are understood to be a general framework encompassing prior and emerging mechanisms in this domain. Whilst the exchange of end-results among cooperatively sensing nodes has been explored in the wireless communication domain and the joint learning via exchange of states has been known in the machine learning community, no viable framework is available to date which quantifies the gains of a docitive system operating in a wireless setting. Numerous problems hence remain, some of which are listed below:

- *Information theory.* One of the core problems is how to quantify the degree of intelligence of a cognitive algorithm. With this information at hand, intelligence gradients can be established where docition should primarily happen along the strongest gradient. This would also allow one to quantify the tradeoff between providing docitive information versus the cost to deliver it via the wireless interface. Some other pertinent questions are how much information should be taught, can it be encoded such that learning radios with differing degrees of intelligence can profit from a single multicast transmission, how much feedback is needed, how often should be taught, etc?
- *Wireless channel.* Whilst not vital to the operation of docitive engines, it is of importance to quantify the coherence times of the wireless medium. This, in turn, allows one to estimate whether the channel dynamics allows for sufficient time for the knowledge dissemination and propagation.
- *PHY layer.* At this layer, as well as all OSI layers above, a pertinent question is which of the states should be learned individually, and which are advantageously taught? Another open issue is how much rate/energy should go into docition versus cognition?
- *MAC layer.* Open challenges relate to the problematic of optimal broad/multicast protocols which allow a single docitive radio to disseminate to as many as possible cognitive entities, all of which could have a different degree of intelligence.
- *Docitive system.* At system level, numerous questions remain open, such as what is the optimal ratio of docitive versus cognitive entities; what is the optimal docition schedule; should every cognitive entity also be a docitive one; what is the docition overhead versus the cognitive gains; etc.
- *Distributed learning.* More specifically to docition, scalability is a problem for many learning techniques and especially for multi-agent learning. The dimension of the search space grows rapidly with the number and complexity of agent behaviors, the number of agents involved and the size of the network of interactions among them. In addition to that, multi-agent systems are typically dynamic environments where the agents learn and the adaptation to one another changes the

environment itself. For this co-adaptation of learners the literature has recently focused on demonstrating the achievement of suboptimal Nash equilibriums, but the convergence to optima is still a wide-open issue.

We believe that we just touched the tip of an iceberg as preliminary investigations have shown that docitive networks are a true facilitator for utmost efficient utilization of scarce resources and thus an enabler for emerging as well as unprecedented wireless applications.

5 Conclusions

Cognitive and cooperative wireless networks are addressed in this section, providing a view on efficient ways of setting networks.

When addressing Cognitive Radio Networks, several aspects of spectrum usage and management are discussed. A number of techniques to be developed for the implementation of efficient spectrum usage through cognitive radio networks are dealt with: spectrum sensing, spectrum management, spectrum mobility, and spectrum sharing mechanisms.

Cognitive Positioning is then addressed, in relation with cognitive radio, Multi Carrier systems being taken as an example. Signal-intrinsic capability for localisation is discussed, namely on the criteria to optimise the function of ranging, and on the characteristics of the signals.

A brief discussion on cooperative networks follows. Colocation of base stations and MIMO relay schemes are among the topics listed in the subsection.

Finally, the concept of Docitive Radios & Networks is introduced, i.e., a novel framework on radios and networks that teach other radios and networks. These radios and networks are not (only) supposed to teach them the end-result, but rather elements of the methods to getting there. A taxonomy is presented, together with an example, a vision, and challenges.

References

1. Sanders F.H.: Broadband spectrum surveys in Denver, CO, San Diego, CA, and Los Angeles, CA: Methodology, analysis, and comparative results. In: Proc. IEEE Int. Symp. Electromagnetic Compatibility, Rome, Italy (1998).
2. McHenry M.A., Tenhula P.A., McCloskey D., Roberson D.A., Hood C.S.: Chicago spectrum occupancy measurements & analysis and a long-term studies proposal. In: Int. W. Technology and Policy for Accessing Spectrum, Boston, MA (2006).
3. Petrin A., Steffes P.G.: Analysis and comparison of spectrum measurements performed in urban and rural areas to determine the total amount of spectrum usage. In: Proc. Int. Symp. Advanced Radio Technologies, Boulder, CO (2005).
4. Chiang R.I.C., Rowe G.B., Sowerby K.W.: A quantitative analysis of spectral occupancy measurements for cognitive radio. In: Proc. IEEE Veh. Technol. Conf., Dublin, Ireland (2007).
5. Wellens M., Wu J., Mähönen P.: Evaluation of spectrum occupancy in indoor and outdoor scenario in the context of cognitive radio. In: Proc. Int. Conf. Cognitive Radio Oriented Wireless Networks and Communications, Orlando, FL (2007).

6. López-Benítez M., Casadevall F., Umbert A., Pérez-Romero J., Hachemani R., Palicot J., Moy C.: Spectral occupation measurements and blind standard recognition sensor for cognitive radio networks. In: Proc. Int. Conf. Cognitive Radio Oriented Wireless Networks and Communications, Hannover, Germany (2009).
7. Federal Communications Commission (FCC), Notice of Proposed Rule Making, ET Docket no. 04-113 (2004).
8. Mitola J. III, Maguire G.Q.: Cognitive radio: Making software radios more personal. IEEE Pers. Commun. **6**(4), pp. 13–18 (1999).
9. Akyildiz I.F., Lee W.-Y., Vuran M.C., Mohanty S.: Next generation/dynamic spectrum access/cognitive radio wireless networks: A survey. Comput. Networks **50**(13), pp. 2127–2159 (2006).
10. Bourse D., et al.: The E2R II Flexible Spectrum Management (FSM) – Framework and Cognitive Pilot Channel (CPC) Concept – Technical and Business Analysis and Recommendations. E2R II White Paper (2007).
11. Alberty T.: Frequency domain interpretation of the Cramér-Rao bound for carrier and clock synchronization. IEEE Pers. Commun. **43**(2-3-4), pp. 1185–1191 (1995).
12. Celebi H., Arslan H.: Cognitive positioning systems. IEEE Trans. Wireless Commun. **6**(12), pp. 4475–4483 (2007).
13. Cramér H.: Mathematical methods of statistics. Princeton Univ. Press, Princeton, NJ (1946).
14. D'Andrea A.N., Mengali U., Reggiannini R.: The modified Cramér-Rao bound and its application to synchronization problems. IEEE Pers. Commun. **42**(2-4), pp. 1391–1399 (1994).
15. Kay S.M.: Fundamentals of statistical signal processing: Estimation theory. Prentice-Hall, Englewood Cliffs, NJ (1993).
16. Mallinckrodt A., Sollenberger T.: Optimum pulse-time determination. IRE Transactions PGIT-3, pp. 151–159 (1954).
17. Mengali U., D'Andrea A.N.: Synchronization Techniques for Digital Receivers. Plenum Press, New York, NY (1997).
18. Skolnik M.: Introduction to radar systems. McGraw-Hill, New York, NY (1980).
19. Van Trees H.L.: Detection, Estimation and Modulation Theory. Wiley, New York, NY (1968).
20. Lottici V., Luise M., Saccomando C., Spalla F.: Non-data aided timing recovery for filter-bank multicarrier wireless communications. IEEE Trans. Commun. **54**(11), pp. 4365–4375 (2006).
21. Luise M., Zanier F.: Multicarrier signals: A natural enabler for cognitive positioning systems. In: Proc. Int. Workshop on Multi-Carrier Systems & Solutions (MC-SS), Herrsching, Germany (2009).
22. (About.com definitions) Cognition – What is cognition? [Online]. Available: http://psychology.about.com/od/cindex/g/def_cognition.htm
23. Private communication with Apostolos Kontouris, Orange Labs, France (2007).
24. Sycara K.P.: Multiagent systems. AI Magazine **19**(2), pp. 79–92 (1998).
25. Galindo-Serrano A., Giupponi L.: Aggregated interference control for cognitive radio networks based on multi-agent learning. In: Proc. Int. Conf. Cognitive Radio Oriented Wireless Networks and Communications, Hannover, Germany (2009).
26. Bellman R.: Dynamic Programming. Princeton Univ. Press, Princeton, NJ (1957).
27. Harmon M.E., Harmon S.S.: Reinforcement learning: A tutorial (2000).
28. Watkins J., Dayan P.: Technical note: Q-learning. Machine Learning **8**, pp. 279–292 (1992).
29. Hoen P., Tuyls K.: Analyzing multi-agent reinforcement learning using evolutionary dynamics. In: Proc. European Conference on Machine Learning (ECML), Pisa, Italy (2004).
30. Tan M.: Multi-agent reinforcement learning: Independent vs. cooperative learning. In: Huhns, M.N., Singh M.P. (Eds.): Reading in Agents. Morgan Kauffman, San Francisco, CA, pp. 487–494 (1993).

The ultimate limits of wireless networks

Carles Antón-Haro (✉), Yonina Eldar, Laura Galluccio, George Iosifidis, Marios Kountouris, Iordanis Koutsopoulos, Javier Matamoros, Sergio Palazzo, Shlomo Shamai, Marc Belleville, Venkatesh Ramakrishnan, Guido Masera, and Dominique Morche

1 Introduction

This paper discusses the grand challenges associated with the holy grail of understanding and reaching the ultimate performance limits of wireless networks. Specifically, the next goal in the networking community is to realize the Future Internet. In this paper, we take a step further from the state of the art in this field, and, describe the main challenges associated with desirable optimal network operation and control in future wireless networks.

First, we start with the challenges associated with max-weight and backpressure type of network control strategies, which have been shown to be throughput optimal. We identify as grand challenges the decentralized light-weight implementation of these algorithms, as well as the issue of coping with flow-level and other types of dynamics. Understanding the deep structural properties of these policies and shaping them towards achieving the goals above in the presence of numerous resource and interference limitations in wireless networks would be an important step forward.

Next, we dwell into game theoretic modeling as a means for understanding and predicting stable network operating points, namely points from which no node has an incentive to deviate. To this end, understanding interaction and convergence to equilibrium points is very important. Furthermore, the community should focus on the notions of competition and cooperation. For the former, non-cooperative interaction would be the best way to model the network, while for the latter, distributed optimization approaches would be desirable. We also put forward mechanism design as a means for enforcing certain desirable equilibrium points. Finally, we also suggest that the network operation under limited and sometimes inaccurate network state information dictates that nodes apply certain learning rules to gradually perceive one's surroundings and readjust their actions.

The next thrust we concentrate on is network coding. Network coding has emerged as a revolutionary paradigm for information transfer in uni-cast or multi-cast traffic. Network coding turns out to be particularly effective and robust in environments of intermittent connectivity and continuous volatility. Network coding brings along

Benedetto S., Correia L.M., Luise M. (eds.): The Newcom++ Vision Book. Perspectives of Research on Wireless Communications in Europe.
DOI 10.1007/978-88-470-1983-6_5, © Springer-Verlag Italia 2012

a number of yet to be resolved challenges in resource efficient network operation; namely coding (traffic mixing) across the network, security and optimal resource utilization.

Stochastic geometry is then considered as an enabler for addressing fundamental performance limits of massive and dense wireless networks through network information theory. The main facets of innovation here entail the understanding of modeling certain spatiotemporal node scenarios and protocol interactions with certain stochastic processes and probability distributions and assess their impact on analyzing network performance.

Sensor networks emerge as a particular class of wireless networks, expected to proliferate in envisioned sensory networks. The main reason of deployment of sensor networks is the estimation of unknown parameters in an area. Key research issues include the sensor network control to fulfill estimation functionalities, realize opportunistic communication and networking under limited spectrum and energy budget. Finally, the notion of constrained cooperation is put forward as means for modeling a wide range of wireless network topologies, infrastructure-based or infrastructureless.

2 Backpressure inspired network control strategies for reaching the ultimate performance limits of wireless networks

2.1 Introduction

The ever increasing demand for efficient and ubiquitous networking poses significant challenges for the wireless networking research community. It is evident that the design of scalable throughput-maximizing resource allocation algorithms is of paramount importance for the efficient operation of large-scale wireless and wireline networks. Early proposed solutions such as random access methods for the access control layer, as well as contemporary legacy access control solutions do not ensure optimal operation and therefore are not sufficient for the emerging traffic-demanding and bandwidth-hungry applications. The backpressure and Max-weight-type class of scheduling algorithms provide a provably efficient mechanism for achieving maximum throughput and guaranteeing queue stability in a wide range of scenarios and seem to be the appropriate direction forward to achieve the holy grail of optimal throughput and quality of service control in wireless networks.

Wireless, perhaps the most rapidly developing segment of the Future Internet today is also the least understood in terms of coping with frequent volatility and extremely dynamic channel conditions. Furthermore, wireless communication is significantly resource constrained since it often has to occur under severe spectrum and energy limitation.

The class of back-pressure and max-weight algorithms was introduced in early 90s in the seminal paper of Tassiulas & Ephremides [1] which presented the MaxWeight scheduling policy for throughput maximization in multi-hop wireless networks. The main idea is to optimally select the subsets of links to be simultaneously activated and to decide how much traffic to route over each activated link. In a subsequent work [2], the same authors described a max-weight policy for allocating a server among several parallel queues with time-varying connectivity. Broadening the latter framework, max-weight type policies were developed for power control and scheduling of shared wireless downlink channels with rate variations, or for joint congestion control, routing and scheduling.

The powerful properties of max-weight type of policies have emerged in the realm of cross-layer control in wireless networks. The reader is referred to Georgiadis et al. [3] and also Srikant [57] for two comprehensive surveys. The distinguishing characteristic of max-weight class of policies is that the subset of nodes that are simultaneously served is selected so as to be of maximum sum 'weight', hence the term 'max-weight'. The weight of a node is usually defined to be the instantaneous backlog (namely, the number of packets in the queue of a node awaiting transmission), or the product of the backlog and the feasible instantaneous service rate for that queue, if selected. In a multi-hop setting, the notion of backlog differential is typically used, i.e., the difference in backlogs with each downstream queue, giving rise to so-called back-pressure mechanisms. The subsets of queues which can be scheduled simultaneously are subject to certain constraints based for example on interference. In a more general sense, max-weight policies can be interpreted as selecting a service vector from a (possibly time-varying) feasible rate region that maximizes the inner product with the backlog vector. Under mild assumptions, Max-weight type algorithms have been shown to provide maximum throughput, i.e., achieve queue stability whenever feasible to do so at all. A particularly appealing feature is that max-weight policies only need information on the current backlogs and instantaneous service rates, and do not rely on any explicit knowledge of the rate distributions or the traffic demand parameters. Another interesting feature is that backpressure routing can be implemented in a distributed fashion since it relies on differential backlog information which can be readily obtained through low overhead communication with a node's neighbours.

2.2 Challenges of backpressure policies

We now present some challenges associated with backpressure style network control strategies.

- *Dynamic Queues.* A fundamental premise is that max-weight policies have been initially developed for systems consisting of a fixed set of queues with stationary, ergodic traffic processes. In real life scenarios, the collection of active queues dynamically varies, as sessions eventually end, while new sessions occasionally start. In many situations the assumption of a fixed set of queues is still a reasonable modelling assumption, since scheduling actions and packet-level queue dynamics tend to occur on a very fast time scale, on which the population of active sessions evolves only slowly. In other cases, however, sessions may be relatively short-lived, and the above time scale separation argument does not apply. The impact of flow-level dynamics over longer time scales is particularly

relevant in assessing stability properties, as the notion of stability only has strict meaning over infinite time horizons. Motivated by the above observations, the stability properties of max-weight scheduling and routing policies in a scenario with flow-level dynamics have been examined, and was proved that max-weight policies may fail to provide maximum stability in that situation [4].
The intuitive explanation is that max-weight policies tend to give preferential treatment to flows with large backlogs, even when their service rates are not particularly favourable, and thus rate variations of flows with smaller backlogs are not fully exploited. Note that the preferential treatment in fact also applies in the absence of any flow-level dynamics. In that case the phenomenon cannot persist however, since the flows with smaller backlogs will build larger queues and gradually start receiving more service, creating a counteracting force. In contrast, in the presence of flow-level dynamics, max-weight policies may constantly get diverted to arriving flows, while neglecting the rate variations of a persistently growing number of flows in progress with relatively small remaining backlogs, so the opposing effect is never triggered.
The design and analysis of suitable scheduling algorithms which guarantee maximum stability in the presence of flow dynamics remains a challenging subject for further research. In future ubiquitous wireless connectivity network scenarios, node mobility will be one of the prevalent characteristics of envisioned networks, and therefore it is of crucial importance to design suitable scheduling algorithms which guarantee maximum stability in the presence of highly dynamic data flows.

- *Decentralized Implementation.* Another key feature related to the dynamics and autonomic operation of future networks, is their decentralized architecture. In max-weight policies, the scheduling and resource allocation mechanisms are not amenable to distributed implementation because every node should be aware of all the queue lengths and the topology state variables in order to independently optimize its own strategy. Furthermore, even if the required information is available, the scheduling might become extremely complicated especially for large networks. Hence, it is evident that distributed implementations of optimum throughput techniques will be one of the main challenges of future networks. The maximum matching derivation is the issue that mostly challenges the distributed implementation of max-weight and backpressure routing and scheduling policies. For max-weight policies, many proposed solutions consider the option of imperfect scheduling at the expense of the feasible capacity region in order to derive distributed low complexity algorithms. In detail, solutions that address the distributed resource allocation with respect to stability can be classified in two major categories. In the first class, there exist algorithms which adopt simplified physical and MAC layer protocols and propose suboptimal scheduling schemes. A basic concern in this case is to bound efficiency loss, [5–7]. For example, in [5], the authors present a simple scheduling policy, referred to as maximal scheduling, which ensures activation of all the non-interfering links. They consider simple interference models and prove that the efficiency loss due to the "distributisation" of the algorithm is bounded. That is, the algorithm guarantees a certain fraction of the optimal centralized solution, which varies with the specific communication and interference model. The second class of methods attempt to derive distributed rate allocation and scheduling algorithms for implementing the max-weight policy without efficiency loss, i.e. they preserve throughput optimality of the centralized solution. This is easily achievable in networks where interference is limited in specific subsets of the nodes. For example, in [7], the authors present distributed algorithms for networks where each node is able to communicate only with a specific subset of nodes, due for example to power transmission constraints. It is assumed that all nodes that belong to the same subset are aware of the link and channel condition and are able to calculate the induced interference. Actually, this is akin to cell partitioned network model where different cells do not interfere with each other, and hence the scheduling can be decentralized to each cell. Networks with orthogonal coding schemes, spatially multiplexing techniques or simply consisting of far located subset of nodes, fall within this class of networks. Nevertheless the problem of efficient and fully distributed implementation of max-weight policies is an important open problem the solution of which is highly related to achieving the holy grail of throughput optimal network control.

2.3 Backpressure policies for single and multi-hop packet switches

The real life implementation of max-weight type of policies today seems more likely to be accomplished than ever. Besides ongoing research in wireless networks, this type of algorithms is used in wire-line networks or in the programming of IP router algorithms in single packet switches. In these switches, the packets waiting to be transmitted are queued in separate queues according to their destinations. This queuing system is usually referred as "Virtual Output Queuing" (VOQ) and in the context of switching fabrics, the architecture based on VOQ and a crossbar (running without internal speedups) is denoted as "pure input-queued (IQ) switch". The max-weight policy is applied by using the VOQ or the waiting time of packets and achieves optimal performance, [8]. However, in the case of interconnected switches, this result is not straightforward. Namely, it was shown in [9] that a specific network of IQ switches implementing a max-weight scheduling policy can exhibit an unstable behavior even when none of the switches are overloaded.

This counterintuitive result opened new perspectives in the research on IQ switches, reducing the value of most of the results obtained for switches in isolation. The challenge here is to leverage research results on the thrust of network switch scheduling to boost transport capacity in wireless networks. Network switches present a simplified scenario, devoid of e.g. interference. Hence, results in this realm can be used as starting points to build up algorithms with desirable properties for wireless networks, as described above.

2.4 Future grand challenges

The future of wireless networking is currently getting shaped by two driving forces: First, the increasing proliferation of communication devices and the development of novel traffic-demanding applications such as VoIP (Voice over IP) and VoD (Video on Demand). It is already evident that these developments will create the need for reliable ubiquitous wireless networking capable of supporting the timely exchange of large amounts of application data. Moreover, following the current trend of users for autonomous operation not relying in central controllers/administrators, designers of future networks should strive to develop distributed protocols that ensure the efficient operation of the communicating nodes.

Therefore, it is very important to enhance the max-weight based policies in order to suit the environment of future communication networks. The two grand challenges are:

- the fully distributed implementation of the backpressure algorithm;
- the support of time sensitive data transfers in a highly dynamic environment.

In future networks, operation under central controllers or administrators is neither feasible nor desirable due to the growing interest of users for unrestricted autonomic communication. Hence, the scheduling component of the max-weight algorithm which requires central and complex computation must be modified. Moreover, the distributed execution should not come at the expense of the network capacity region, a requirement which still remains unsatisfied. Creating variants of the backpressure and max-weight algorithms that are amenable to light-weight, low-overhead implementation is then of paramount importance.

Additionally, node mobility and other dynamic effects turn out to degrade the algorithm performance or even un-stabilize the network. In future networks these characteristics will be even more prevalent and therefore any promising resource allocation policy should take them into account. Moreover, since nodes will be operating independently and protocol compliance is not a priori given, all policies should be designed in order to deter selfish or even malicious behaviour. In particular, since max-weight and backpressure strategies will rely on decentralized information exchange with neighbours, it is very challenging to design techniques that induce truthful declaration of certain node quantities (e.g. queue lengths) in order to guarantee desirable execution of these algorithms.

3 Game theoretic modelling and understanding of node interactions in future autonomic networks

3.1 Introduction

In envisioned future networks where the autonomic operation and the complex interactions of nodes are expected to be the prevalent characteristics, game theory should be employed both as a performance analysis and a protocol design tool. In these systems, whether they are ad hoc wireless networks, peer-to-peer overlays or online-offline social networks, the nodes operate independently and, moreover, their operation is constrained by their limited resources, such as battery life, link capacity, connection time duration, etc.

With the proliferation of communication devices and the development of novel traffic demanding applications, these issues will become even more complicated and profound. In case all nodes belong to and are controlled by a single entity, as for example in a military deployable ad hoc network, the above issues are translated into a problem of coordination. On the other hand, when the nodes act independently, the issue of lack of cooperation also arises. Therefore, we anticipate that in the future there will be an extreme need both for coordination achieving schemes and cooperation enforcement mechanisms as a prerequisite for efficient network operation.

3.2 Coordination and competition problems

Starting from the coordination problem, one should devise appropriate distributed algorithms to guarantee efficient network resource allocation. Efficiency is captured by certain network specific metrics such as achievable throughput rate and maximum node lifetime or, more generally, by the abstract network utility maximization metric as discussed for example in [10] and [3]. The employed algorithms should be lightweight, with low communication overhead and loose synchronization requirements in order to be practical and realizable [11].

Relaxing the assumption of node cooperation, game theory appears to be the most suitable mathematic tool for the study of these networks. Traditionally used by economists to analyze strategic agents' interactions [12], it has recently been employed as a means for capturing the inherent competition dynamics in autonomous networks, ranging from the future Internet [13], up to wireless ad hoc networks [14]. Game theory can be used to study node behaviour, enable the prediction of their interaction outcome, and therefore facilitate the design of respective protocols. As we move towards more ubiquitous and autonomous modes of network operation, *non-cooperative* game theory is expected to be used for modelling the nodes strategies, so as to capture how the actions of each node affect the utility of others and predict the reached equilibriums. Moreover, cooperative game theory will be used for the analysis of systems in which the nodes initially form groups and cooperate in order to complete a certain task and in the sequel compete for the apportion of the obtained resource or utility as, for example, in [15].

3.3 Learning in games

Node interactions in a dynamic and constantly changing environment dictate that every node can observe past outcomes, update its strategy and improve future actions. Hence, the issue of nodes learning models arises naturally. Natural questions seek an answer in that setting, such as: in which way should one node consider its experience from past interactions in devising his new – current strategy? Is it important to observe other nodes actions and relate them with their gains – reimbursements? What is the impact of limited information, namely when the node is not aware of all the other nodes actions and results? There exist various methods that model the learning process and evolution, which vary in required assumptions and properties. In the future, these issues have to be addressed in a more systematic fashion by combining results from different disciplines such as cognitive science [16], learning theory [17], and game theory [18]. In particular, it is of paramount importance to study the impact of different learning schemes onto the result of nodes interaction in the presence of limited information, noise or even malicious agents and nodes who attempt to disrupt the coordination-cooperation procedure.

3.4 Mechanism design

Once the nodal interaction and learning machinery is modelled and properly understood, the next step is to devise efficient mechanisms for ensuring the efficient network operation. First, the method of pricing which has already been extensively applied both for wireless [19] and wire-line [20] networks is a basic tool to control the nodes behaviour and drive their interaction in desired equilibrium points. However, for autonomous systems, and due to the lack of information about nodes preferences and utilities, auction mechanisms appear to be more suitable. In this case, the charging prices are not predetermined, rather are defined according to the submitted bids. There exist several results about the efficiency of these algorithms/mechanisms, see for example [21], and it is crucial to study how we can cast them for the autonomous systems of interest. For example, it is important to devise methods for implementing auctions without requiring charging infrastructure, [22], based solely on service reciprocation among nodes.

3.5 Future challenges

In light of emerging future autonomous networks, game theory based analysis and modelling should be enhanced in order to satisfy long cultivated anticipation of research community. First, it is necessary to derive more detailed models that capture all the aspects of the novel communication paradigms and therefore result to more stringent conclusions and realizable protocols. In this context, one should expect models which do not preclude the strategy space or the rules of nodes interaction. On the contrary, the possible actions of each player – node, the stages and the repetitions of interactions, the utility functions and other components of the game should be identified dynamically.

Additionally, in future large and decentralized networks each node will operate with limited information about other nodes and the network state. In this setting, it is interesting to study possible equilibrium points and their respective properties such as efficiency loss, computational complexity, etc. Thus, understanding the repercussions of various types of learning algorithms and the way they impact convergence to equilibrium points should be properly addressed.

Moreover, the large number of nodes – players yields for the employment of evolutionary game theory where nodes are grouped with respect to their behavioural profile. By observing the results of the repeated interactions, it is possible to identify the dominant strategies and therefore derive significant insights to be used in the efficient protocol design.

4 Network coding

Recently, the emerging paradigm of *network coding* has received much attention. Network coding is a new research area that promises interesting applications not only in information and coding theory, but in practical networking systems too [23]. In traditional networks, coding is used for two different purposes. At source nodes, source coding is applied for data compression and therefore reduction of the required transmission bandwidth. On the other hand, channel coding is used at the link level to ensure reliable communication, thus enabling to model links as essentially error-free channels subject to the channel capacity. Only recently, in the paper by Ahlswede et al. [24], it was shown that, in general, to restrict coding to these two applications is not optimal. Even in a scenario where all redundancy at the sources has been removed and channels are error-free, encoding data streams at intermediate nodes in general increases the capacity of the network. The improvement is most pronounced when network traffic volume is near the maximum capacity obtainable with traditional routing [25].

In contrast to traditional network approaches that merely relay and replicate data at intermediate nodes, network coding calls for nodes to perform algebraic operations on the received data. Though similar in spirit to source coding in its use of encoded source data, the fact that network nodes can actively operate on the data, gives network coding additional and surprisingly powerful capabilities. Drawing on these capabilities, network coding has the potential to significantly reduce delay and power consumption while improving bandwidth utilization and overall network robustness for a diverse set of applications across a wide variety of network technologies, such as overlay, wireless, and sensor networks. These applications include unicast transfers of data along multiple paths, reliable multicast, and best-effort dissemination of control traffic.

Moreover, another interesting benefit of network coding might be in terms of robustness and adaptability. Intuitively, network coding, similarly to traditional coding, takes information packets and produces sets of encoded packets, where each of them is "equally important". Provided that the destination receives a sufficient number of encoded packets, no matter which, it is able to decode and obtain the information required.

4.1 Future challenges and open issues in network coding

In the near future, with the high increase in the use of wireless communications, network coding will assume a significant role. Indeed, wireless communications, based on broadcast transmissions, provide the natural setting to exploit the benefits of network coding. Nevertheless, it should be taken into account that applying network coding in wireless and sensor networks introduces implementation problems different from the wired network setting. The most salient aspect of wireless transmission is the use of broadcast via omnidirectional antennas. Broadcasting may provide beneficial collateral transmission to neighbouring nodes for free, but also causes interference. The use of network coding may significantly alter the nature of the interplay between advantages and drawbacks. Another key example is the communication/computation tradeoff of using network coding within sensor networks. Since communication is relatively expensive compared to computation in sensor networks, network coding may offer substantial advantages in power consumption.

Moreover, network coding will have an emerging role within the context of *cooperative networking*. The aim of cooperation is that network communications become more reliable and efficient when nodes "support" each other to transmit data. The cooperation can be achieved by enabling neighbouring nodes to share their own resources and their power with the hope that such a cooperative approach leads to savings for both the overall network resources and power consumption. Network coding should be thought of as a form of user cooperation, where the nodes not only share their resources and their power but also their computation capabilities.

As an example, it has been demonstrated in [26], by using both simulation studies and realistic experiments, that network coding in Avalanche may improve the overall performance of *peer-to-peer* content distribution, up to about a hundred of peers. The intuition that supports such a claim is that, with network coding, all blocks are treated equally, without the need to distribute the "rarest block" first, or to find them in the "end game" of the downloading process.

In the peer-to-peer networks of the future, network coding approaches are expected to minimize download times. Because the performance of the system depends on the specific overlay topology and schedule, with network coding approaches, very simple mechanisms that construct a random overlay can be used. Moreover, due to the diversity of the coded blocks, a network coding based solution is much more robust in case the server leaves the network early and, in contrast to forwarding based protocols, network coding protocols suffer only a small performance penalty when incentive mechanisms to cooperate are implemented.

Another field of future application for network coding is the emerging *extreme networking*, i.e. networks characterized by intermittent connectivity, where the construction of a continuous end-to-end path between source and destination is difficult or impossible. In this kind of networks, a network coding based solution allows to disseminate information to all nodes in the network with a high probability at a significantly lower overhead and with no help of routing protocols. Moreover, one interesting characteristic of the network coding approach in this kind of networks is that it benefits more from node mobility than other approaches. Furthermore, even for extreme conditions such as a sparse mobile network with a high packet drop rate and with nodes that sleep for a considerable amount of time, the network coding approach performs reasonably well, whereas a protocol based on classical routing is simply unusable.

Though there are already a large number of scenarios where network coding has proven its merit, we believe that this list is far from being exhausted. In fact, new applications of network coding are reported almost every month. Some of these applications are theoretically well understood, whereas in some cases the performance gains are difficult to analyze mathematically and have to be determined by means of simulations and measurements.

In spite of the large body of research results on network coding, there is still a gap to be bridged between theory and practice, before a broad deployment of this technology is conceivable. We believe that in the near future it should be determined what impediments need to be overcome and what are the true benefits and the underlying tradeoffs when using network coding. The main reasons today against the development of practical network coding is that an immediate killer application has not yet been identified, and the gains, while non-trivial, are not enough to cover many of the inherent problems (network support, backward compatibility, etc.) in deploying network coding.

For example, many network coding strategies are not yet fully deployable due to a high computational complexity or a significant decoding delay. The development of low complexity architectures, even suboptimal ones but with practical engineering advantages is one of the main challenges that will be faced. Another question that impacts the utility of network coding is the amount of computation and buffering it requires from intermediate nodes. Although analytical results in the network coding literature focus on what is possible theoretically, and although there are some initial results that bound the resources required, experimentation is necessary to determine actual performance in real systems.

4.2 *Optimization of network coding functionalities in future networks*

Multi-session network coding is another issue expected to apply in future networks. In particular, the general problem of multi-session network coding, where multiple communicating sessions with independent data share a network of lossless links with rate constraints, has proven to be extremely challenging. In single-session network coding, every receiver wants the same sources of information, hence it is not too surprising that randomly mixing the packets in the network is an asymptotically optimal strategy [27]. In contrast, the generalization to the case of multiple multicast sessions remains open. In multi-session network coding, if packets are randomly mixed in the network, a receiver may receive many coded packets but may still not be able to decode them since they are mixed with unwanted packets. In general, combining more packets together increases the diversity of the composition, as it may potentially be useful to more receivers, however, this makes more difficult for receivers to clean out the pollution from a mixture packet. Intuitively, it is the need for careful encoding and decoding of packets that makes this problem hard.

From a theoretical point of view, some recent studies have characterized the capacity region for general acyclic multi-source multi-sink networks with arbitrary transmission, closing the gap between the existing inner and outer bounds [28]. However, how to explicitly evaluate the obtained capacity region still remains an open problem.

Besides, *resource optimization* is crucial. With classical network coding techniques, the output packets are generated combining the buffered packets by using a set of randomly generated coefficients. The problem is how fast output packets should be generated. This can be formulated as a mathematical optimization problem, where the variables are the allocated link rates and the sending rate at the source. Several recent studies have proposed centralized and distributed, primal and dual algorithms for the optimization, but a lightweight near optimal approach is still desired.

Finally, *security* is yet another significant active area of research in the field of network coding. For example the *pollution attack*, where a malicious node injects a junk packet into the network, is far more destructive if network coding is used than in a traditional system. If the junk packet is stored into the buffer of a node, the buffer will be polluted, the output of the node will become junk, and this may soon propagate to the entire network. In the near future, some counter-measures should definitely be implemented and tested.

5 Fundamental limits of wireless networks using stochastic geometry

5.1 Introduction

Developing a general end-to-end capacity theory for wireless networks is the holy grail of network information theory. Despite the significant progress made in recent years, a holistic and concise integration of networking and information theory (IT) remains elusive, largely due to difficulties in extending mutual information concepts to multiuser network problems. Shannon IT has established the essential underpinnings of modern communication theory, by providing fundamentals limits and design insights for links and configurations with very few nodes. However, classical IT seems to be far from sufficient for characterizing the ultimate limits of large random networks (e.g. MANETs), maintaining the networking and IT union unconsummated.

Several recent approaches have tackled the problem of developing a unified network theory, ranging from capacity scaling laws for dense random or extended networks and cut set bounds for erasure networks to the deterministic channel model and degrees of freedom capacity approximations. Despite providing significant insight on the large-scale performance, most of the existing theoretical tools fall short of providing the fundamental limits. A finer view of throughput limits is needed, which will embrace the random and highly dynamic nature of networks, particularly the stochastic interactions among nodes at different timescales. For that, tools from stochastic geometry [29] and random graphs [30] appear to be promising and powerful pathways to attack longstanding challenges and circumvent current methodological insufficiencies.

5.2 Relevance of stochastic geometry and random graphs

Stochastic geometry is a rich branch of applied probability that allows the study of physical phenomena characterized by highly localized events distributed randomly in a continuum. Intrinsically related to the theory of point processes [31, 32], it has come into prominence in recent years with a plethora of applications in fields ranging from biology, geography, and astronomy to operations research, image analysis and communication engineering. Although stochastic geometry has been used to characterize interference in wireless networks at early 70's, its use for modeling communication networks is relatively new yet rapidly increasing.

A wireless spatial network can be viewed as a set of interacting points-nodes interconnected by communication paths; as such, at a given time, a collection of nodes transmit concurrently towards their respective receiver(s) creating interference. In this context, the two main performance limiting factors, namely received signal intensity and interference, both present a dominant spatially dependent component, which has to be understood and taken into account during the network design process. As signal to interference plus noise ratio (SINR) becomes the relevant figure of merit, stochastic geometry appears to be particularly relevant for capturing the macroscopic properties of stationary spatial networks and quantifying – among others – interference, connectivity, outage probability, and information progress. When ergodicity can be established, these properties are abstracted by averaging over potential node geometrical distributions within a given parametric class.

This leads the definition of various spatiotemporal averages that capture the key dependencies of the network performance characteristics as functions of a relatively small number of network operating parameters. In other words, the network components (nodes, communication links, service zones) are represented as a family of random objects (point patterns, graphs, tessellations) and relevant network performance characteristics are then expressed as functional of stochastic processes that depends only on their distribution parameters.

5.3 Current and emerging research approaches

The use of spatial models in wireless communications dates back to the late 70's [33,34], but a series of papers beginning in the late 90's revitalized the interest in modeling and analyzing wireless networks using stochastic geometry [35,36]. Several relevant metrics, i.e. the expected information progress and the transmission capacity [37], have been recently introduced, finding application in designing and analyzing the performance of a wide variety of wireless ad hoc networks, including spread spectrum, interference cancellation, spectrum sharing, overlaid and cognitive radio networks, scheduling and power control, and multi-antenna techniques (MIMO). The interested reader is referred to several very helpful recent tutorials and monographs [38–41].

The state of the art contributions can be categorized into qualitative results on infinite population models, where conditions and/or scaling laws for network connectivity, percolation, and packet average delivery delay are provided; and quantitative results, which provide closed-form expressions for both time and space averages assuming a certain physical layer model, medium access control (MAC) and routing protocols. The vast majority of recent studies adopt the key assumption that the positions of the transmitting nodes (interferers) form a stationary Poisson point pro cess (PPP) on an infinite plane. The Poisson distribution is often a reasonable model for networks with uncoordinated transmissions (e.g. ALOHA, random access MAC protocols) or substantial mobility. Note that the homogeneous PPP, whose entropy is maximum among all stationary point process with the same intensity, constitutes a worst-case distribution in many aspects and has the benefit of being analytically tractable.

5.4 Short-term research directions

As mentioned earlier, the majority of current results assume a (stationary) homogeneous Poisson field of interferers mainly for analytical tractability. Nevertheless, practically relevant network settings require going beyond the Poisson point process assumption and dealing with in-homogeneity. This is a key theoretical enabler in order to model and analyze existing and emerging physical layer transmission techniques and intelligent medium access and routing schemes. Performing smart local scheduling will result in non homogeneous Poisson transmitter locations, while idealized centralized scheduling can eliminate outages altogether and determine the optimal set of transmitters in each slot (e.g., Max-Weight scheduling within the back-pressure paradigm).

A first problem to study is whether a CSMA/CA scheme or a channel-aware extension of ALOHA will provide gains over uncoordinated transmissions. Preliminary work in this direction employed either a Matern hardcore process or the Simple Sequential Inhibition point process to model CSMA/CA-based networks and compute the outage probability and transmission capacity. Furthermore, it is of significant relevance to investigate wireless network settings with clustered (due to MAC-induced or geographical clustering), and regular (e.g. lattice point process) node spatial distributions [42].

Modelling non-homogeneous geometrical distributions is not an easy task; adequate and tractable heterogeneous point process models and analytical techniques are yet to be identified in the near future. Poisson driven, Neyman-Scott, and Thomas cluster point processes are of potential interest as they still allow for many explicit formulas, while offering enhanced flexibility in modeling than the class of homogeneous PPP. Going one step further, in order to overcome the limitation of the excessively deterministic character of the PPP intensity function, Cox process models seem attractive as a means to model aggregated or clustered point patterns. Briefly, Cox point processes (or doubly stochastic PPP) are a direct generalization of the mixed Poisson process, obtained by randomizing the intensity measure in a PPP. Successful results towards this direction will allow us to design dense and self-organizing wireless networks with enhanced spatial reuse and/or coverage. Furthermore, cutting-edge radio technologies, such as MIMO systems and interference alignment, combined with network coding can also be incorporated in a framework that embraces heterogeneity. Moreover, in upcoming studies, secrecy constraints, multi-hop packet forwarding and coexistence issues (guaranteed QoS in overlaid/multi-tier networks) ought to be incorporated into the stochastic geometric models.

Whenever exact closed-form expressions of relevant performance metrics (e.g. throughput, outage probability, delay, etc.) are hard or impossible to obtain, several efficient bounding and approximation techniques are required to be developed. First, perturbation analysis can be employed to analyze non-homogeneous Poisson models, where an underlying homogeneous Poisson process is perturbed, resulting in a non-homogeneous point process, whose characteristics can be approximated by some expansion (e.g. the Taylor expansion) in the neighborhood of the homogeneous scenario. Second, the theory of stochastic ordering and directionally convex functions may provide elegant and efficient tools for comparison of random measures on locally compact spaces. This will allow us to assess the impact of deterministic displacement of points, independent superposition and thinning and i.i.d. marking on the Palm distribution of point processes.

Hence, such tools will enable us to quantify how certain performance characteristics of wireless networks improve with more variability in the input node distribution process. Another promising tool is that of deterministic space transformations on the first- and second-order characteristics of the underlying point process. This method aims at transforming a non-stationary point process into a weakly stationary, as a means to homogenize the first two moments. Finally, Poisson process approximation for locally dependent point processes and for dependent superposition of point processes using Stein's method, and Berry-Esséen type bounds might also assist towards that direction. It should be noted for some of the above techniques, as well as to derive extreme value results for point processes, that the interconnection of point processes with shot-noise processes is exploited.

5.5 Long-term research directions

The concept of self-organization is central in future wireless networks, especially in cases where there is no medium to exchange data between nodes other than the spontaneous and heterogeneous wireless links between neighboring nodes. This together with the user mobility and the randomness already present in most MAC protocols require a completely new methodology for understanding large-scale, spatio-temporal dynamics within a stochastic geometric framework. Gibbs point processes and the random closed sets that one can associate to them look particularly promising for designing and analyzing distributed measurement based self-organized wireless networks. The main challenge of this approach is how to benefit from the locality of the Gibbs sampler in order to design distributed algorithms that take decisions based on minimal exchanges with neighbors, which also result in a global network optimization. Since Gibbs models are rarely amenable to closed form analysis, at least in the case of systems with randomly located nodes, one should expect that a significant amount of time and effort need to be invested before obtaining fruitful and significant results.

Going beyond specific topologies, developing a general global view concerned with network asymptotics will significantly advance our understanding of the ultimate limits of wireless networks. For that, a variety of simpler network abstractions, such as the radial transmission model (disk graph), and other graph-based, SINR and stochastic geometrical models are to be considered. Such global view and network abstractions will be essential to characterize the throughput-delay-reliability (TDR) surface for large networks using the methods of dynamic and delay-aware information theory.

Furthermore, once the research community obtains a better understanding of two-dimensional, possibly non homogeneous, communication network models, a natural extension of increased complexity is to consider the network variability in time as well as the spatial fluctuations of the traffic. For that, insightful three-dimensional space-time network models have to be developed, together with properly defined time-scale decompositions: depending on the timescale, a single realization of the point process or an ensemble of realizations is relevant, providing thus fundamental "separation theorems" for key network functionality. Moreover, such models may also provide the necessary degrees of freedom to incorporate upper layer processes and parameters (e.g. access rate, traffic, admission control, routing decisions, etc.) into aggregate interference characterization. This will result in more accurate modeling of dynamic interactions and advanced cross-layer techniques, allowing for efficient optimization of network characteristics and protocols (MAC, routing, power control).

Evidently, a sustained, visionary, cross-disciplinary research effort has to be undertaken by the scientific community in order to characterize the fundamental limits of wireless networks. In the distant future, we expect to see an interactive and vivid dialogue of stochastic geometry and random graphs with theories ranging from non-convex optimization theory, stochastic learning and inference (e.g. for self-organizing and/or bio-inspired algorithms), compressive sensing (e.g. for neighbor discovery, random access protocols, spectrum sensing) to statistical mechanics, game theory, and disequilibrium economics. Ultimately, developing a powerful arsenal of advanced, elaborated, and efficient space-time stochastic geometrical models and techniques may achieve multiple goals (politely), thus being applicable in a vast and general category of complex network problems (e.g. social, biological, information, etc) and establishing the first solid theoretical foundations of network science.

6 Sensor networks

6.1 Introduction

Over the last decades, research on Wireless Sensor Networks has attracted considerable attention. Typically, a Wireless Sensor Network is composed of a large number of energy-constrained devices called sensors. Such sensor nodes are commonly equipped with finite batteries and simple radio interfaces to transmit their measurements to a fusion center (FC), where the data is processed. Typical applications in WSNs encompass: environmental monitoring (e.g. temperature, pressure), inventory tracking and military applications, to name a few.

A vision of the future could be a future application of a WSN. Thousands of lightweight sensors are thrown to the atmosphere and dragged by the wind. Sensors collect measurements of a phenomenon of interest, e.g. a toxic cloud, and, subsequently, their readings are transmitted to a more powerful device for data storage and processing. Sensors are equipped with very low batteries which can be recharged via harvesting methods (e.g. wind energy).

Energy harvesting is expected to be a key issue in future WSNs: sensors will be able to recharge their batteries from environmental energy and, thus, contribute to the vision of green wireless communications. Furthermore, not only WSN stand on the basis of green communications, but also they help build more energy efficient and, thus, more environmentally-friendly applications. In this context, a current application that is already possible is the design of smart buildings (e.g. efficient lighting and heating in office buildings).

The possibility to build inexpensive and small sensor nodes with stringent energy budgets opens a new paradigm in research. How should the information be aggregated in order to avoid sending redundant data over the wireless links? How can fading be combated? What are the most suitable topologies for WSNs? In general, the answers to these questions remain open, and strongly depend on the application for which a given WSN is deployed to. Besides, due to the large number of sensor nodes forming the network, centralized signal processing techniques and communications protocols are not suitable for this kind of networks. This follows from the fact that centralized signal processing techniques and communication protocols often entail the acquisition of global information (e.g. instantaneous channel gains, position of the nodes...) which is barely desirable, due to the large amount of signalling that this entails.

6.2 State of the art

As commented above, a typical application of a WSN is that of the *estimation* of an unknown parameter of interest. In this context, sensors observe a common source of interest, of a random nature, in order to produce an estimate of some parameter of interest (e.g. humidity, temperature, etc). For a number of reasons, such as hardware limitations, the observations in the sensors are affected by some measurement noise. Such noisy observations are then transmitted to the FC where the parameter of interest associated to the underlying source is estimated. This is the so-called CEO (Chief Executive/Estimation Officer) which was introduced by Berger [43] for which, the rate distortion function for the quadratic Gaussian case was recently derived in [44].

The source-channel coding separation theorem, by which source and channel coding can be regarded as decoupled problems and thus be solved independently [45], turns out to provide suboptimal solutions in the case of Multiple Access (MAC) channels with correlated sources. In [46], the authors prove that an amplify-and-forward (A&F) strategy is known to scale optimally, in terms of estimation distortion, when the number of sensor nodes grows without bound. However, such asymptotic optimality is achieved if distributed synchronization of the sensor signals can be orchestrated at the physical layer to achieve beam-forming gains, which is barely achieved in a WSN. Conversely, for *orthogonal* channels and, thus, pair wise synchronization for each sensor-to-FC communication, solutions based on the source-coding separation theorem outperform A&F-based ones. Indeed, in [47] the source-coding theorem is shown to be optimal for the quadratic Gaussian CEO problem.

As far as source coding is concerned, the work in [48] constitutes a generalization to sensor trees of Wyner-Ziv's pioneering studies [49]. More precisely, the authors compare two different coding schemes, namely Quantize-and-Estimate (Q&E) and Compress-and-Estimate (C&E). The former is a particularization of Wyner-Ziv's problem to the case where no side information is available at the decoder; whereas the latter is a *successive* Wyner-Ziv-based coding strategy capable of exploiting the correlation among sensor observations. Yet sub-optimal, the performance of this successive encoding scheme is not far from that of other more complex joint (over sensors) coding strategies. The most important aspect here is that sensors distributedly compress their information by using statistical information about the side information already available at the FC. In both cases, it is assumed that sensors experience *Gaussian* (or, more generally, *deterministic*) channels, i.e. the transmissions rates are known and, consequently, all sensors convey their observations *reliably*.

6.3 Future challenges

In this section we state a number of challenges associated with sensor networks.

- *Spectrum management.* It is projected that in the next 10–15 years, wireless sensors networks will be an important part of our lives. That is, hundreds of thousands, if not millions of small sensor nodes will surround us and form different wireless sensor networks. In a wireless sensor network (e.g. for parameter estimation) it is of paramount importance to ensure a predefined estimation quality on the parameter of interest. For this reason, one of the main issues of future WSNs is its operation in a crowded spectrum scenario where multiple WSNs, all of them application-specific, have to coexist.
- *Cross-layer designs (MAC/PHY).* In [48], the authors address the problem from an information-theoretical perspective. In particular, it is assumed that each sensor has a reserved and orthogonal channel for its communication or, in other words, sensors operate under a centralized reservation-based MAC protocol (e.g. TDMA/FDMA). Unfortunately, central coordination/scheduling of the sensor-to-FC transmissions entail extensive signalling which is barely desirable in WSNs. For that reason, an attractive approach is to address a cross-layer design encompassing the MAC *and* PHY layers by adopting realistic multiple access schemes such (CSMA or, ALOHA). In these cases, the probability of collision turns out to be directly related to the metric of interest in our WSN.
- *Opportunistic Communications.* A fundamental characteristic of the wireless channel is the fluctuation of the channel strength due to constructive and destructive interference between multi-paths. This fluctuation, known as fading, can be combated by creating different independent paths between the transmitter and the receiver and as a result, obtaining *diversity*. In a multipoint-to-point network one can exploit the fact that different users act as different antennas obtaining the so-called *multi-user diversity* (MUD). The exploitation of MUD in WSNs has been

recently introduced in [50] for decentralized parameter estimation. The authors consider two problems of interest: (1) the minimization of distortion subject to a sum-power constraint, and (2) the minimization of transmit power subject to a maximum distortion target. In both cases, the optimal power allocation is given by a kind of water-filling solution, in which sensors with poor channel gains or noisy observations should remain inactive to save power. However, a water-filling like solution entails global CSI at the FC to compute the optimal power allocation. For this setting, a simple and low feedback strategy is proposed in [51] to reduce the feedback load and still retain most of the optimality of the water-filling like solution.

- *Network topologies.* The most popular topology studied in the literature is the many-to-one sensor network [50], i.e. a given number of sensors attempting to transmit their data to the FC. Unfortunately, sensors are low-powered and long-range transmissions might be unrealistic. In this situation, one can instead deploy a heterogonous WSN, which is composed of sensors with different capabilities. For instance, sensors can be organized into clusters, being each cluster under the supervision of a more powerful device called cluster-head (CH). Hence, each CH is now in charge of consolidating the cluster estimate and sends the information to the FC. Clearly, one could also adopt this network topology with homogeneous sensors by selecting a sensor as a cluster-head depending on the current network conditions (e.g. the one with the strongest channel gain to the FC or, the one with the higher residual energy). In all cases, an important open issue is the processing of the data at the cluster-head. Various options are possible: e.g. consolidation of the cluster information into a cluster estimate and re-transmission to the FC, ii) re-transmission of all the sensors measurements to the FC or, ii) hybrid strategies.

7 Constrained cooperation in wireless networks

By now it is evident that future wireless technology and in particular multi-cell based system requires cooperation on different levels, as to be able to efficiently use the potential available and to come close to the information theoretic ultimate limits. Here we shortly point to work, partly done under NEWCOM++ support that highlights those aspects, and provides not only bounds, but in fact sheds lights on novel methodologies and approaches that have the potential to come close to ultimate performance in practice.

We provide here a short representative list referencing our work, giving the flavor of the techniques, and methodologies used, calling attention to the information theoretic perspective. We trust that these methodologies will play role in future wireless technologies.

Specifically, we mostly focus on the performance of network Multi-Input-Multi-Output (MIMO), also referred to as Multi-Cell Processing (MCP), but we also consider the interplay of such technique with cooperation in the form of relaying at the user terminals [52–55].

The impact of future technologies as highlighted by information theory is best viewed by noticing how fast theoretical studies became the centre of practical interest, and a classical example is multi-user detection, as well as the capacity of the MIMO broadcast channels. In the later, paradigms, like dirty-paper coding were involved.

The references here mainly address simplified cellular models that extend the Wyner-type models considered in the initial works of the early nineties. Specifically, the studies consider the presence of limited-capacity and limited-connectivity backhaul links, fading channels and the interplay of MCP with relaying.

7.1 Future prospects

The massive body of theoretical work exemplifies the advantages of future wireless technologies, which will be based on far reaching processing abilities, harnessing thus the theoretical perspective into practical approaches. Cooperation among the base stations and occasional relays has been shown to be able to potentially increase the capacity of the network by an amount proportional to the inter-cell interference span (i.e., number of base stations interfered by a local transmission) with respect to standard single-cell strategies with spatial reuse.

While initial work demonstrated such benefits under idealistic conditions, in terms of, e.g., absence of fading and perfect backhaul, more recent research has confirmed the promises of MCP under more practical conditions. The performance benefits of cooperation at the terminal level have been addressed as well, along with considerations regarding the strong interplay between the design of relaying strategies and of MCP techniques. Our studies have also briefly touched upon the potential gains achievable by exploiting novel transmission strategies such as structured codes. Other advanced techniques, such as interference alignment, cross-layer design and robust coding, are also expected to have an important role to play in cooperative future wireless technologies, and we expect the latter to impact on the more remote time horizon.

8 Conclusions

In this paper, we have presented the main facets of research challenges that lie ahead towards the goal of understanding the ultimate performance limits of networks, and of designing innovative techniques to approximate and even achieve them.

Various optimization- and control theory driven techniques where put forward such as distributed optimization and the max-weight control principle. Novel and disruptive approaches were described such as network coding, and their potential was analyzed. Furthermore, mathematical tools such as cooperative and non-cooperative game theory, learning theory and stochastic geometry will be needed in order to model and understand the spontaneous interactions of massively dense autonomic networks. This paper presented a holistic approach with techniques that are amenable to realization in various classes of wireless networks, from sensor networks to infrastructure-based and infrastructure-less wireless networks.

Clearly, the way ahead is promising. Novel and disruptive approaches will need to be undertaken, oftentimes relying on cross-disciplinary techniques migrating from a wide range of disciplines and thrusts as described above.

References

1. Tassiulas L., Ephremides A.: Stability properties of constrained queueing systems and scheduling policies for maximum throughput in multihop radio networks. IEEE Trans. Aut. Contr. **37**(12), pp. 1936–1948 (2009).
2. Tassiulas L., Ephremides A.: Dynamic server allocation to parallel queues with randomly varying connectivity. IEEE Trans. Information Theory **39**(2), pp. 466–478 (1993).
3. Georgiadis L., Neely M.J., Tassiulas L.: Resource allocation and cross-layer control in wireless networks. Foundations and Trends in Networking **1**(1), pp. 1–144 (2006).
4. van de Ven P.M., Borst S.C., Shneer S.: Instability of MaxWeight scheduling algorithms. In: Proc. Conference on Computer Communications, Rio de Janeiro, Brazil (2009).
5. Lin X., Shroff N.: The impact of imperfect scheduling on cross-layer rate control in multihop wireless networks. In: Proc. Conference on Computer Communications, Miami, FL (2005).
6. Neely M.J., Modiano E., Li C.: Fairness and optimal stochastic control for heterogeneous networks. In: Proc. Conference on Computer Communications, Miami, FL (2005).
7. Chaporkar P., Kar K., Sarkar S.: Throughput guarantees through maximal scheduling in wireless networks. In: Proc. Allerton Conf. Control, Commun., Comput., pp. 1557–1567, Monticello, IL (2005).
8. McKeown N., Mekkittikul A., Anantharam V., Walrand J.: Achieving 100% throughput in an input-queued switch. IEEE Trans. Commun., **47**(8), pp. 1260–1267 (1999).
9. Andrews M., Zhang L.: Achieving stability in networks of input-queued switches. In: Proc. Conference on Computer Communications, Anchorage, AK, (2001).
10. Chiang M.: Balancing transport and physical layers in wireless multihop networks: Jointly optimal congestion control and power control. IEEE J. Sel. Areas Commun. **23**(1), pp. 104–116 (2005).
11. Bertsekas D., Tsitsiklis J.N.: Parallel and Distributed Computation: Numerical Methods. Prentice-Hall, Upper Saddle River, NJ (1989).
12. Fudenberg D., Tirole J.: Game Theory. MIT Press, Boston, MA (1991).
13. Papadimitriou C.: Game theory, algorithms, and the Internet. In: Proc. ACM-SIAM Symposium on Discrete Algorithms, Washington, DC (2001).
14. Theodorakopoulos G., Baras J.S.: A game for ad hoc network connectivity in the presence of malicious users. In: Proc. Global Telecommun. Conf., San Francisco, CA (2006).
15. Mathur S., Sankaranarayanan L., Mandayam N.B.: Coalitional games in Gaussian interference channels. In: Proc. IEEE Int. Symposium on Information Theory (ISIT), Seattle, WA (2006).
16. Dayan P., Daw N.D.: Decision theory, reinforcement learning, and the brain. Cognitive, Affective & Behavioral Neuroscience, pp. 429–453 (2008).
17. Awerbuch B., Kleinberg R.: Competitive collaborative learning. Journal of Computer and System Sciences **74**, pp. 1271–1288 (2008).
18. Fudenberg D., Levine D.: The Theory of Learning in Games. MIT Press, Boston, MA (1998).
19. Xi Y., Yeh E.M.: Pricing, competition, and routing for selfish and strategic nodes in multi-hop relay networks. In: Proc. Conference on Computer Communications, Phoenix, AZ (2008).
20. Cao X.R., Shen H., Milito R., Wirth P.: Internet pricing with a game theoretical approach: Concepts and examples. IEEE/ACM Trans. Networking **10**, pp. 208–216 (2002).
21. Nisan N., Roughgarden T., Tardos E., Vazirani V.V.: Algorithmic Game Theory. Cambridge Univ. Press, Cambdrige, UK (2007).

22. Iosifidis G., Koutsopoulos I.: Double auction mechanisms for resource allocation in autonomous networks. IEEE J. Sel. Areas Commun. **28**(1), pp. 95–102 (2010).
23. Katti S., Rahul H., Hu W., Katabi D., Medard M., Crowcroft J.: XORs in the air: Practical wireless network coding. IEEE/ACM Trans. Networking **16**(3), pp. 497–510 (2008).
24. Ahlswede R., Cai N., Li S.R., Yeung R.W.: Network information flow. IEEE Trans. Information Theory **46**(4), pp. 1204–1216 (2000).
25. Liu J., Goeckel D., Towsley D.: Bounds of the gain of network coding and broadcasting in wireless networks. In: Proc. Conference on Computer Communications, Anchorage, AK (2001).
26. Gkantsidis C., Rodriguez P.: Network coding for large scale content distribution. In: Proc. Conference on Computer Communications, Miami, FL (2005).
27. Jaggi S., Sanders P., Chou P.A., Effros M., Egner S., Jain K., Tolhuizen L.: Polynomial time algorithms for multicast network code construction. IEEE Trans. Information Theory **51**(6), pp. 1973–1982 (2005).
28. Yan X., Yeung R.W., Zhang Z.: The capacity region for multi-source multi-sink network coding. In: Proc. IEEE International Symposium on Information Theory, Nice, France, (2007).
29. Stoyan D., Kendall W., Mecke J.: Stochastic Geometry and Its Applications, 2nd ed. Wiley and Sons, New York, NY (1996).
30. Penrose M.: Random Geometric Graphs, Oxford Studies in Probability. Oxford University Press, Oxford, UK (2003).
31. Kingman J.: Poisson Processes. Oxford University Press, Oxford, UK (1993).
32. Daley D., Jones D.V.: An introduction to the theory of point processes. Springer, New York, NY (1988).
33. Kleinrock L., Silvester J.A.: Optimum transmission radii in packet radio networks or why six is a magic number. In: Proc. National Telecommunications Conference (1978).
34. Musa S., Wasylkiwskyj W.: Co-channel interference of spread spectrum systems in a multiple user environment. IEEE Trans. Communications **26**(10), pp. 1405–1413 (1978).
35. Baccelli F., Klein M., Lebourges M., Zuyev S.: Stochastic geometry and architecture of communication networks. J. Telecommunication Systems **7**(1), pp. 209–227 (1997).
36. Baccelli F., Zuyev S.: Poisson-Voronoi spanning trees with applications to the optimization of communication networks. Operations Research **47**(4), pp. 619–631 (1999).
37. Weber S., Yang X., Andrews J., de Veciana G.: Transmission capacity of wireless ad hoc networks with outage constraints. IEEE Trans Information Theory **51**(12), pp. 4091–4102 (2005).
38. Baccelli F., Blaszczyszyn B.: Stochastic geometry and wireless networks. Foundations and Trends in Networking (2009).
39. Haenggi M., Andrews J.G., Baccelli F., Dousse O., Franceschetti M.: Stochastic geometry and random graphs for the analysis and design of wireless networks. IEEE J. Sel.Areas Commun. **27**(7), pp. 1029–1046 (2009).
40. Win M., Pinto P., Shepp L.: A mathematical theory of network interference and its applications. Proc. IEEE **97**(2), pp. 205–230 (2009).
41. Weber S., Andrews J.G., Jindal N.: A tutorial on transmission capacity. Submitted to IEEE Trans. Commun. (2009). [Online]. Available: http://arxiv.org/abs/0809.0016
42. Ganti R.K., Haenggi M.: Interference and outage in clustered wireless ad hoc networks. IEEE Trans. Information Theory **55**, pp. 4067–4086 (2009).
43. Berger T., Zhen Z., Viswanathan H.: The CEO problem: Multiterminal source coding. IEEE Trans. Information Theory **42**(3), pp. 887–902 (1996).
44. Prabhakaran V., Tse D., Ramchandran K.: Rate region of the quadratic gaussian CEO problem. In: Proc. IEEE Int. Symposium on Information Theory (ISIT), Chicago, IL (2004).
45. Cover T.M., Thomas J.A.: Elements of Information Theory. Wiley Series in Telecommunications, New York, NY (1993).
46. Gastpar M., Vetterli M.: Source-channel communication in sensor networks. Lecture Notes in Computer Science, pp. 162–177 (2003).
47. Xiao J.-J., Luo Z.-Q.: Multiterminal source-channel communication over an orthogonal multiple-access channel. IEEE Trans. Information Theory **53**(9), pp. 3255–3264 (2007).
48. Draper S.C., Wornell G.: Side information aware coding strategies for sensor networks. IEEE J. Sel. Areas Communications **22**(6), pp. 1–11 (2004).

49. Wyner D., Ziv J.: The rate-distortion function for source coding with side information at the decoder. IEEE Trans. Information Theory **22**(1), pp. 1–10 (1976).
50. Cui S., Xiao J., Goldsmith A., Luo Z.-Q., Poor H.V.: Estimation diversity and energy efficiency in distributed sensing. IEEE Trans. Signal Process. **55**(9), pp. 4683–4695 (2007).
51. Matamoros J., Antón-Haro C.: Opportunistic power allocation schemes for wireless sensor networks. In: Proc. IEEE Symposium on Signal Processing and Information Technology, Cairo, Egypt (2007).
52. Levy N., Zeitouni O., Shamai S. (Shitz): Central limit theorem and large deviations of the fading Wyner cellular model via product of random matrices theory. Problems of Information Transmission **45**(1), pp. 5–22 (2009).
53. Simeone O., Somekh O., Poor H.V., Shamai S. (Shitz): Local base station cooperation via finite-capacity links for the uplink of wireless networks. IEEE Trans. Information Theory **55**(1), pp. 190–204 (2009).
54. Levy N., Zeitouni O., Shamai S. (Shitz): Information theoretic aspects of users' activity in a Wyner-like cellular model. IEEE Trans. Information Theory **56**(5), pp. 2241–2248 (2010).
55. Somekh O., Simeone O., Poor H.V., Shamai S. (Shitz): Cellular systems with full-duplex amplify-and-forward relaying and cooperative base stations. In: Proc. IEEE Int. Symp. Inform. Theory, Nice, France (2007).
56. Simeone O., Somekh O., Kramer G., Poor H.V., Shamai S. (Shitz): Uplink sum-rate analysis of a multicell system with feedback. In: Proc. Allerton Conference on Communication, Control, and Computing, Monticello, IL (2008).
57. Shakkottai S., Srikant R.: Network Optimization and Control. Foundations and Trends in Networking **2**(3) (2007).

Bandwidth and energy efficient radio access

Andreas Polydoros (✉), Hanna Bogucka, Piotr Tyczka, and
Carles Navarro Manchon

Within the last two decades, the amount and diversity of services provided by wireless systems has been drastically transformed. Mobile (cellular) communication, for instance, is nowadays offering a wide variety of multimedia-data services, in contrast to the limited voice and very simple data services offered in the past. In wireless local-area networks (WLAN), as another example, the ability to be on-line without needing a wired connection is not sufficient any more, and users expect to experience similar data speeds and quality of service (QoS) as with a wired connection. This has lead to a rapid increase in data-rate requirements ("broadband connectivity") in the standards of new and upcoming wireless communication systems.

In order to increase the data rates offered, a simple approach would be to increase the bandwidth allocated to a certain system. However, the proliferation of applications that use the air interface as the transmission medium limits the amount of available bandwidth in the radio-frequency spectrum, making it a very scarce and costly resource. Therefore, research over the last years has been focused towards improving the *spectral efficiency* of wireless communications, so that higher data rates can be achieved within a given bandwidth. For example, the deployment of multiple transmit antennas and their stream-multiplexing ability (MIMO), along with the requisite advances in receiver architectures, has been a critical step in achieving this goal.

This paper discusses briefly the current theoretical as well as practical research advances that address the problem of bandwidth- and energy-efficient radio access and also provides a brief description of yet-unsolved but interesting problems for future research in this area. The first part of this paper discusses briefly the theoretical background of the problem and then proceeds with a discussion of various techniques that are useful in achieving the theoretical limits, namely (1) advanced coding/decoding techniques; (2) Adaptive Modulation and Coding (AMC) and (3) cognitive spectral usage. Specifically, a more detailed look at coding/decoding techniques and, in particular, iterative techniques for wireless receivers will be the topic of the second part of this paper. The last part will discuss cognitive spectrum usage (always for the purpose of advancing spectral utilization) with a special focus on applied game theory for this goal.

Benedetto S., Correia L.M., Luise M. (eds.): The Newcom++ Vision Book. Perspectives of Research on Wireless Communications in Europe.
DOI 10.1007/978-88-470-1983-6_6,

1 Fundamental communication limits in fading

The purpose of this section is to provide a brief survey of schemes that model the existence of multiplicative noise (fading) in a wireless environment, since such fading is the main source of channel variability that degrades the quality of broadband connections. Our main goal here is neither to provide an exhaustive analysis of the topic, nor to explain in depth the existing techniques. Instead, we aim to show the importance of the work in the topic and emphasize the directions for future work.

1.1 Channel capacity

The evaluation of capacity bounds for various channel models and system scenarios has been an interesting research topic since Shannon's pioneering work. The determination of such performance limits defines the framework for the design and development of optimal communication techniques. In many cases, the computation of the exact value of channel capacity is a difficult problem and bounding tehcniques are needed. In some cases, these bounds can lead to the exact determination of channel capacity. In all cases however, any available analytical characterization of a system information capacity serves as a performance criterion and a tool for the system design. The ergodic (Shannon) capacity was derived under the assumption of codes with infinite codeword length, extending over different realizations of the fading state. It represents the expected value of the capacity, averaged over the different channel fading states. Practical systems employ codes with length much shorter than the channel coherence time. In such systems the distribution of the instantaneous capacity is needed for performance characterization. In most cases the outage capacity is used as a performance metric, defined as the capacity that can be guaranteed for a percentage value of time.

1.1.1 Channel capacity of fading channels with channel side information

In wireless transmission channel conditions may vary arbitrarily due to changes in the fading environment or due to the users' mobility. In most practical cases, and depending on the relative dynamics, the system is able to extract valuable information regarding the prevailing level (strength) of the channel-fading process. This type of information is usually referred to as CSI and can either be available at both ends (the transmitter as well as the receiver), or at the receiver (Rx) only but not at the transmitter (Tx). The degree of accuracy of this CSI is certainly a critical parameter for the performance of the system under consideration. The case in which the channel is assumed to be perfectly known at the receiver and/or the transmitter has been studied extensively in the literature. In a Rayleigh-fading Single-Input Single-Output (SISO) setting, it was addressed in an early work by Ericsson [1], where analytical expressions for the capacity of flat-fading channels with perfect receiver CSI have been derived. In a more recent work (Ozarow et al. [2]), results were derived for the average as well as the *outage* capacity in cellular mobile radio, assuming perfect CSI at the Rx.

In recent OFDM-type designs, techniques are implemented whereby the estimated values of the CSI are relayed to the Tx with different levels of accuracy. It has therefore been of particular interest to re-visit earlier results on the capacity of fading channels but now with Tx-side information available. Goldsmith and Varaiya [3] analyzed the capacity of flat fading channels with perfect CSI at the transmitter and/or the receiver. Borade and Zheng [4] investigated, among other scenarios, the channel capacity in the low signal-to-noise-ratio (SNR) regime when both sides have perfect CSI. They proved that, for very low values of SNR, the capacity is SNR log (1/SNR) and this is achieved by on-off signalling with a fixed "on" level.

In practice, true access to perfect CSI cannot be guaranteed to either side, mainly due to the rapid changes in the fading environment and limited energy that can be allocated to non-data carrying symbols to assist the channel estimation. Thus, extensive research has addressed the case where CSI is either imperfect or, in the other extreme, completely unavailable to either side. Abou-Faycal et al. [5] have studied the capacity of the Rayleigh fading channel with purely unknown fading levels. They showed that the optimal input distribution is discrete. This is a somewhat surprising result, especially when compared with the optimality of the continuous Gaussian input distribution when CSI is available and perfect. Under a peak constraint on the input of a Rician fading channel, this discreteness property of the capacity-achieving input distribution has also been proven in [6–8] and, more generally, for a broad class of SISO channels in [9]. In a MIMO setting the capacity and optimal input distribution of a Rayleigh fading channel have been derived by Marzetta and Hochwald [10] where CSI is also unavailable to both sides. This work triggered further research on the discreteness of optimal input distributions on MIMO settings. Zheng and Tse [11] addressed the capacity of MIMO Rayleigh fading channels in high SNR.

In most cases of interest, however, these extremes, namely of either having perfect CSI or no CSI at all, are not valid. In particular, practical OFDM systems tend to be between those two extremes, except when there is very high mobility. The analysis of the capacity of fading channels with imperfect or "noisy" CSI is therefore of great practical interest. Medard [12] investigated the effect of imperfect channel knowledge on the channel capacity and obtained upper and lower bounds on the achievable mutual information rates. Lapidoth and Shamai [13] analyzed the effects of channel estimation errors on performance whenever Gaussian codebooks are employed along with nearest-neighbour decoding. The capacity of imperfectly known fading channels is addressed in [14] for the low-SNR regime and in [15] for the high-SNR regime. These results, however, have not considered explicit training and estimation techniques and the needed resources allocated to do so. An analysis of channel capacity in a training-based communication setting with Rayleigh block-fading can be found in [16]. Hassibi and Hochwald [17] looked into some training schemes for multiple-antenna channels recently.

Regarding fading-channel capacity when CSI (i.e., the channel fading value) is available at the Tx, the CSI is treated as causal side information at the Tx (as opposed to non-causal side information that is not treated in this survey). In this case, techniques such as adaptive rate/power control, MIMO beam-forming, water-filling etc. are all applicable. The causal case of Tx-CSI was first introduced by Shannon [18] wherein he showed that the CSI-endowed channel can be transformed into a no-CSI channel of an exponentially larger alphabet size. Assuming such causal Tx-CSI, we now describe briefly the results derived for typical fading models in the literature.

- *Slow fading.* When the Tx knows the CSI, one option is to control the transmit power such that the corresponding full information rate can be delivered regardless of the fading state. Regarding the power this effectively "channel inversion" strategy guarantees a constant receiver SNR, irrespective of the channel gain. Regarding the rate adaptation some pre-specified rates are chosen when such exact channel inversion is feasible. However, a very large amount of power must be spent in order to 'invert' a very bad channel, which encounters practical limitations of peak-power-constrained transmission.
- *Fast Fading.* The goal now is to maximize the *average* information rate where the averaging occurs over many coherence-time periods. The optimal power and rate allocation in this case is based on the water-filling principle. In general, the Tx allocates more power when the channel is good, taking advantage of this improved channel condition, and less or even nothing when the channel is poor. This is conceptually the reverse of the channel-inversion strategy above. The natural implication of the water-filling capacity is a variable-rate coding scheme.

In the above setting, water-filling is done over time. A duality exists with a frequency-selective channel, where water-filling is done over the OFDM subcarriers. In both cases, the problem can be viewed as that of a bit-power allocation scheme over parallel channels.

Comparing the channel capacity in the case of full CSI at the Tx with the one in the Receiver-Only CSI (RxO-CSI) case, some conclusions can be drawn. In particular, at low SNR, the capacity with full CSI is significantly larger than the RxO-CSI capacity, whereas at high SNR the difference between the two tends to zero. Over a wide range of SNR, the gain of the water-filling procedure over the RxO-CSI capacity is very small. A comprehensive survey of information theoretic results on fading channels can be found in [19]. With imperfect channel knowledge at the transmitter, the capacity is $\beta \mathrm{SNR} \log(1/\mathrm{SNR})$, where β is a scalar parameter ($0 \leq \beta \leq 1$), describing the fraction of channel energy in the part of the channel known to the transmitter. An analysis of partial transmitter knowledge over Rician channels can also be found in [6,7].

1.1.2 MIMO-channel capacity

In the above setting, the introduction of multiple antennas under suitable conditions provides an additional spatial dimension for communication and yields gains in degrees of freedom which result in an increase in capacity: In fact, the capacity of such MIMO channels with N transmit and receive antennas is proportional to the number N. MIMO communication is a broad and interesting topic with many applications. In particular, in the high-SNR regime, MIMO techniques become the primary tools to increase capacity significantly through the degree-of-freedom gain previously mentioned, as well as the induced power gain.

The use of multiple transmit and receive antennas provides many benefits for both fast and slow fading channels. In fast fading, antenna diversity introduces a power gain as well as a degree-of-freedom gain. The analysis of fast-fading MIMO is simpler and addresses mainly channel capacity, whereas the analysis of slow fading is generally more complex. In this case, the outage probability is the proper metric to look at as a function of the targeted transmission rate. In slow fading, there is a triple gain from the introduction of multiple antennas, namely in power, degrees of freedom and diversity. In the high-SNR regime, there is an approximation of the outage probability that captures the benefits of MIMO communication for slow fading channels [20]. This is the fundamental trade-off between the increased data rate (via an increase in the spatial degrees of freedom – the multiplexing gain) and the increased reliability (via an increase in the diversity gain). The optimal diversity-multiplexing trade-off is typically used as a benchmark in comparing the various space-time schemes and is helpful for the design of optimal space-time codes.

Considering the time invariant Gaussian MIMO channel, the spatial dimension plays the same role as the time and frequency dimensions in the time-varying fading channel with full CSI and the time-invariant frequency-selective channel. The capacity is therefore obtained by a water-filling power allocation scheme, albeit the water-filling takes place in the spatial domain. It depends highly on the singular values of the channel gain matrix, corresponding to the eigen-modes of the channel (the "eigen-channels"). For high SNR, where the level of "water" is low, it is asymptotically optimal to allocate equal amounts of power on the nonzero eigen-modes. The number of spatial degrees of freedom represents the dimension of the transmitted signal as modified by the MIMO channel. It provides a crude measure of the capacity of the channel. At a low SNR, the optimal policy is to allocate power only to the strongest eigen-mode. In this regime, the rank of the channel matrix is less relevant for the characterization of the channel capacity. Instead, the energy transmitted through the channel is a more critical parameter. For a detailed analysis of MIMO communication as well as the related concepts the reader is referred to [20].

1.2 Communication techniques approaching the limits

When the channel fading level is known at the transmitter, then the Shannon capacity is achieved by adapting the transmit power, data rate, and coding scheme relative to this known fading level [3]. As mentioned before, in such a fading environment, AMC is a powerful class of techniques for improving the energy efficiency and increasing the data rate over the fading channel. Therefore, codes designed for AWGN channels can be combined with adaptive modulation for fading channels, with the same approximate coding gains. In what follows we provide a brief discussion of coding/decoding techniques first (the following section), and then we comment on AMC.

1.2.1 Improving the existing coding and decoding techniques

In general, coding techniques for the fading channel should be constructed by taking into account the distinctive features of the underlying model. There, one relevant question is how code design for the fading channel differs from that for the AWGN channel when we assume, for example, a flat, slowly-fading channel plus AWGN. The quest for optimal coding schemes in such a fading environment has led to the development of new criteria for code design. If the channel model is not stationary, as it happens in a mobile-radio system, then a code designed for a fixed channel will in all likelihood perform poorly when the channel varies. Therefore, a code optimal for the AWGN channel may actually be suboptimal for a substantial fraction of time. In these conditions, antenna diversity with maximal-gain combining may prove indispensible: in fact, under fairly general conditions, a channel affected by fading can be turned into an effectively AWGN channel by increasing the number of diversity branches. Another robust solution is based on bit interleaving, which yields a large diversity gain due to the choice of powerful convolutional codes coupled with a good bit interleaver and the use of a suitable bit metric.

The construction of optimal coding techniques is an area of intense scientific interest. Coding and decoding has already played and will continue to play a central role in future wireless systems. Apart from improving the theory and practice of LDPC/Turbo and related families of codes, extended research activity can be found for the improvement of coding and decoding techniques for convolutional codes also.

For MIMO systems, space-time block codes can realize the full-diversity gain and divide the vector-ML decoding problem into simpler scalar problems, a trick which dramatically reduces receiver complexity. Space-time trellis codes yield better performance than space-time block codes by achieving higher coding gain at the cost of increased receiver complexity. The area of MIMO communication theory is fairly new and full of challenges. Some promising MIMO research areas are: MIMO in combination with OFDM and CDMA, new coding, modulation, and receive algorithms, combinations of space-time coding and spatial multiplexing, MIMO technology for cellular communications, and proper AMC techniques in the context of MIMO systems.

1.2.2 Adaptive modulation and coding (AMC)

The objective of AMC is to optimize the use of available resources (system bandwidth, channelization, transmit power, time slots, computational power of executing platform) in order to achieve a specific Quality of Service (QoS). The AMC system is composed of: (1) an adaptation criterion or cost function, generally related to the QoS parameters; (2) the hypothesized nature of the CSI that the transmitter needs to know about the channel, which information is often imperfect, erroneous or obsolete; (3) the particular optimization algorithm chosen or developed for this problem; and (4) the resulting outputs of the AMC optimization algorithm (namely, a set of transmission parameters to be employed in the next transmission). As far as the adaptation criteria are concerned, there are multiple possibilities and combinations of AMC schemes in which some particular metrics and quantities may constitute the inputs (requirements and constraints) for one strategy or the criterion/output for the other. Some quantities (such as data throughput) may be introduced as the criterion (objective function), while others (like transmission power limit and delay) may be introduced as constraints. By varying these choices a large number of adaptation

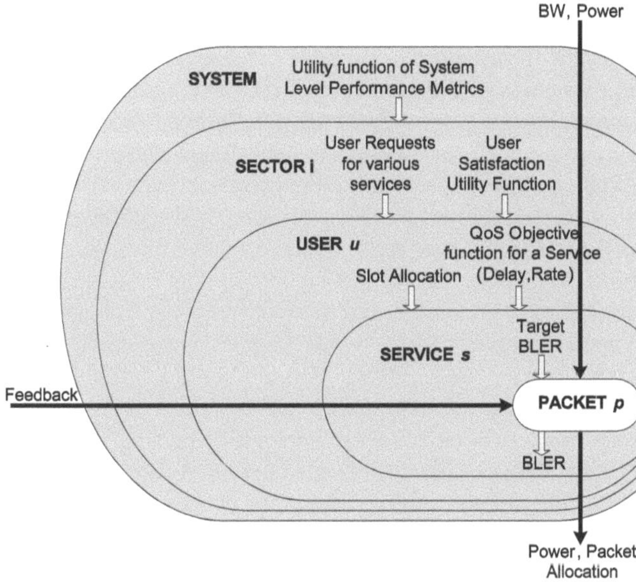

Fig. 1. AMC framework

strategies are obtained. Accommodating that plethora of options is challenging and, in some cases, may result in very complex solutions that are hard to implement.

An illustrative example of possible choices for cross-layer optimization in a cellular system is depicted in Fig. 1. Tx parameter optimization procedures at the system/sector level (higher layers) are usually called *RRM procedures*, while at the user/service layers *AMC algorithms*. All the optimization scenarios are intersected by the two basic common resources, bandwidth and power, and have as input the feedback information (feedback can contain various forms of information, CSI, error measurements, etc.). The objective function used for each layer is usually called the utility function. Utility functions are also used in cross-layer optimization. They are defined so as to balance efficiency and fairness when allocating resources in systems with heterogeneous services. Consequently, they can be used to optimize radio resource allocation for different applications and to build a bridge between the physical, MAC, and higher layers.

At the centre lies the basic element of a packet-based communication system, namely the 'packet'. The interpretation of a packet herein is that of the smallest portion of a communication entity employed by a given system (containing unique information content). It can either be considered correctly received, or discarded at the end of the receiver's processing at the link level, thus determining the performance of the link. This approach is compatible with packet-based systems (like WiMax and LTE); where each packet is characterized by the code rate, block size and constellation used. It can also describe uncoded systems where a symbol can be considered as a packet.

When designing AMC algorithms, the packet Link-Level Performance (LLP) prediction under some CSI information is a fundamental requirement for all optimization problems. In some cases of interest (e.g. coded OFDM-based systems), the exact LLP function is difficult to be derived in an analytic form amenable to run-time optimization. This arise the need for a Compact Link-Level Performance Estimation (CLLPE) model that take into consideration the parameterization of the transmitted signal, given the channel and interference conditions. It should be detailed enough to include channel modelling issues such as: the effect of multiple antennas at the transmitter and/or the receiver, the MIMO technique applied (e.g., beam-forming or other spatial multiplexing scheme), and the receiver type. The accuracy of such models is more crucial in lower layers, while rougher approximations can be used at higher layers.

A review of possible adaptation strategies can be found in [21] with extensive bibliography on the subject, as well as a review on the link adaptation research history. A short summary of this review is presented herein. As mentioned before, there are multiple possible combinations of the adaptation criteria, requirements and constraints. For example, in the throughput-oriented strategy, the AMC algorithm aims in providing the highest bit rate (or spectral efficiency) for a required BER and fixed radiated power limit. This was one of the first adaptive transmission schemes proposed by Steele and Webb [22] for single-carrier QAM modulation and narrow-band fading channels. Exploiting the time-variant channel capacity, various concatenated coded schemes with an adaptive coding rate have been investigated in [23]; variable coding rate and power schemes in [3, 24–27]; latency and interference aspects with turbo-coded adaptation in [28,29]. The concepts elaborated for adaptive QAM modulation and coding have been invoked for OFDM QAM in [30, 31]. Adaptive subcarrier selection for OFDM TDMA dynamic links has been investigated in [32–34], space-time diversity in [35, 36], and multi-coded systems in [37, 38], as well as in the investigation of the key agents affecting AMC performance [39–42]. Interesting proposals for throughput-oriented AMC algorithms can also be found in [43–47].

The set of transmission parameters amenable to adaptation is in general large (i.e. constellation size, code rate, Tx power, symbol rate, number of subcarriers, the number of antennas, etc.). Targeting to reasonable complexity is also a decisive factor for the parameters' selection. For example in [48] it was shown that by adjusting only the power or only the bit rate the resulting capacity is negligibly smaller than by adjusting both these parameters. In [26] a study on maximizing the spectral efficiency by optimally varying combinations of the transmission rate and radiated power with average power and instantaneous (or average) BER constraints has been presented. There, both continuous-rate adaptation and discrete-rate adaptation are considered. The conclusion has been that the use of only one or two degrees of freedom in adaptation yields spectral efficiency close to the maximum possible that would be obtained by utilizing all degrees of freedom.

In energy-constrained (or energy-optimized) wireless networks, a common adjustable parameter is the overall power consumption of a transceiver for a target QoS level. The motivations for this approach are usually the following: to extend the battery life, to minimize the electromagnetic radiation in populated areas, to reduce the cost in infrastructure-based networks and, finally, to reduce the interference level generated. In this case, the transmission parameters should be adapted in order to minimize the power needed for the baseband signal processing and the power of the transmitting antenna. Thus, some trade-offs in choosing the optimization criteria are needed, such as the accuracy of the performance prediction (CLLPE model) within a given limit for the baseband processing. Interesting propositions for power-oriented AMC algorithms and strategies can be found in [24,49–53].

There are interesting proposals in the literature concerning multi-user adaptive schemes. In multi-user systems (for instance, in adaptive FDMA) subcarriers may be adaptively allocated to users in an optimal manner, taking the quality of each user's channel into account. An example is the algorithm described in [54] which aims at minimizing the overall transmit power by adaptive multi-user subcarrier, bit, and power allocation. In [55] a general link and system performance analysis framework is developed that is used to compare the downlink performance of fully loaded cellular system with different types of link adaptation.

2 Iterative techniques in wireless receivers

2.1 Introduction

The discovery of turbo-codes in the early nineties [56] created a revolution in the way wireless receivers were designed. In contrast to the classical structure of sequential receivers, in which parameter estimation, channel equalization and data decoding are performed in an independent fashion, a new generation of receivers based on the turbo-principle started to arise, initially applied to multi-user detection in a CDMA context [57,58] or for turbo-equalization in frequency-selective channels [59], and very successfully extended to MIMO detection and decoding later on. These turbo-receivers are derived heuristically, with each of the receiver blocks (channel estimator, detector and channel decoder) being independently designed and exchanging information between them in an iterative manner. While they were shown to achieve remarkable performance, the exchange of information between different blocks is derived in an intuitive (non-formal) way, making it difficult to find a global design criterion or guarantee their convergence. In order to overcome this drawback, several analytical frameworks for the design of iterative receivers, such as the sum-product algorithm [60] for factor graphs or variational-Bayesian methods [61,62], have been derived. These frameworks, which are often formulated as message passing algorithms in graphical models, aim at the iterative optimization of the estimates of the unknown parameters according to a certain metric or cost function. This provides a global criterion for optimization of the whole receiver design, rather than optimizing each of the receiver modules individually as it is done in the heuristic methods.

The recent advances in iterative processing have allowed the design of iterative receivers achieving close-to-optimal performance under the design assumptions. However, many challenges are still left unsolved: The complexity issues of iterative receivers are one of the most important; due to the iterative nature of the receiver, it is clear that the number of operations to perform in these types of receivers will be significantly larger than in the case of a sequential receiver. Furthermore, due to the iteration with the channel decoder, some latency in the detection of the information bits is introduced, which might be unacceptable for certain kind of applications. In order to solve this, future research will have to focus on the development of low-complexity, high performance iterative solutions, which could be derived from the analytical design frameworks by constraining the set of solutions allowed. Also, research on the field of low-complexity decoding will be crucial for the applicability of iterative receivers in practical solutions, as it will significantly alleviate the computational burden of having to perform several successive decoding iterations of the same signal.

- Most of the iterative solutions designed so far have only been derived under rather ideal conditions and only address questions such as channel estimation, detection and decoding. In the future, the design of iterative structures should also include considerations such as synchronization, RF imperfections (e.g. phase noise) or estimation and mitigation of other system's interference.
- While several analytical frameworks for the design of iterative solutions exist, each fulfilling different optimization criteria, it is not clear which the best approach for a given problem is. Therefore, much of the research effort in the coming years will be put in the analysis and comparison between the different formal approaches, in order to achieve full understanding of their singularities, advantages and drawbacks. Information geometry might be an interesting tool to be used in this context.

In the following, we briefly discuss the different frameworks existing for the design of iterative processing algorithms, and give some examples of their application to iterative receiver design for wireless communications. Finally, some considerations on the important topic of low-complexity decoding algorithms are drawn.

2.2 Framework for iterative processing

As stated above, the first instances of iterative receivers were designed following intuitive criteria. In these solutions, the basic modules of the receiver are kept the same as in the classical sequential receivers, and they exchange information with one another in an iterative fashion. While the performance shown by these type of receivers is indeed outstanding, the lack of analytical justification for their structure makes it difficult to achieve full understanding of their working principles, advantages and limitations. To overcome this drawback, a number of analytical frameworks for the design of iterative processing algorithms have been applied to the problem of iterative receiver design for wireless communications over the last decade. In this section, we review two of them which have attracted a large interest from the research community in the recent years: the sum-product algorithm for factor-graphs and variational-Bayesian inference.

A factor graph [60] is a visual representation of a global function which can be factorized in a certain number of factors. This representation is usually applied to a global probability distribution which can be factorized; in this case, the graph shows the dependency between the different variables. When using such a representation, the sum-product algorithm provides a way of computing the marginals of each of the variables represented in the graph. In this framework, the relation between variables can be interpreted as messages that the variables exchange. By computing and appropriately combining these messages, an estimate of the marginal probabilities of each variable is achieved. The application of the algorithm yields the exact marginals under the assumption that the graph is cycle-free. When the graph has cycles in it, however, the application of the sum-product yields an iterative algorithm which provides approximations of these marginals.

Variational-Bayesian inference [61,62] is another powerful tool for the design of iterative algorithms. In this framework, an auxiliary function with a certain structure is postulated and iteratively updated to approximate the probability distribution of interest. Usually, the auxiliary function is assumed to be factorized with respect to the variables in the model considered, and each of the factors is successively updated by optimizing a given divergence measure of the auxiliary function with respect to the true probability distribution. This divergence measure is usually the Kullback-Leibler divergence [63]. Since the divergence is locally minimized with respect to each of the factors, it is guaranteed that the algorithm yields a sequence of estimated auxiliary functions with decreasing Kullback-Leibler divergence with respect to the true probability distribution.

Recently, several attempts to relate different iterative processing frameworks have been made. In [66], Minka presents a unifying view of different message-passing algorithms, showing that the difference between them is not the amount of structure they model, but the measure of loss they minimize. He generalizes these types of algorithms into a framework in which a generalized divergence measure, the alpha-divergences, is applied to obtain all the different solutions of classical message-passing algorithms. More investigation in this field will help improving the understanding of the research community about these methods, and will help determine which kinds of approaches are preferred for each type of problem.

2.3 Applications on iterative receiver design

The theoretical frameworks presented above have successfully been applied as a solution to many problems in communications. Graphical models have been used to solve the decoding of both LDPC and turbo-codes. For instance, it has been shown that the iterative decoding of turbo-codes can be seen as an instance of the sum-product algorithm [64]. Inspired by the results obtained for decoding, the frameworks have been applied to the design of iterative receivers for wireless communications.

In [65], iterative receivers based on the factor-graph approach are derived for several kind of fading channels. It is shown how restricting the form that the different messages can take can effectively reduce the complexity of the receiver. Also, because the factor-graph approach perfectly defines the message update rules, handling of soft information and proper treatment of extrinsic information are handled automatically. In [67], factor graphs are applied to the problem of multiuser detection in a CDMA system. The factor graph has cycles in it, and therefore different updating schedules of the messages can be applied, yielding different types of receivers. It is also discussed how to approximate messages to obtain low-complexity detectors. Finally, the factor-graph framework was also applied to pilot-assisted interleave-division multiple access (IDMA) systems in [68].

Variational-Bayesian inference methods have also been widely applied to the design of iterative receivers. In [69], this method is applied to data detection and channel estimation in an OFDM system with phase noise. The receiver iterates between the data detector and a Kalman-filter-like channel tracking module. In [63], an instance of these methods, the variational-bayesian expectation-maximization (VBEM) algorithm is applied to MIMO detection and channel estimation in a GSM system. The variational-Bayesian framework is also used in [70] (where it is denominated Divergence Minimization framework), applied to the problem of multi-user detection and channel estimation for a CDMA system. In this work, it is shown how choosing an appropriate factorization of the auxiliary function used can lead to an iterative structure similar to the heuristic turbo-receivers, therefore giving a formal justification for these intuitive structures.

While there exist many examples of applications of these analytical frameworks to the design of iterative wireless receivers, still some issues remain unsolved. The solutions obtained may, in some cases, be still too complex to apply as of now to real communication systems. However, with the recent advances in electronics and computing, and by further investigation in the direction of low-complexity versions of the solutions yield by the frameworks, it is safe to say that iterative receivers will be crucial in achieving the ambitious spectral efficiency requirements that current and future wireless communications systems will impose.

2.4 Low-complexity decoding

The above presented theoretical framework for iterative processing and research directions stemming from that are complemented with low-complexity decoding issues. Since decoding is an inherent part of an iterative receiver, and a task that will be performed more often in an iterative receiver than in a sequential one, the availability of efficient and computationally-reduced decoders becomes crucial for the feasibility of this kind of architectures. Although the topic of simplified algorithms for iterative decoding has already appeared shortly after introduction of turbo codes [56], it still remains live and important issue for both product and LDPC [71] codes.

In iterative scenario, decoding algorithms need to output not only the most likely information bits (or symbols), but also an estimate of the bit reliabilities for further processing. Such soft-output algorithms perform a forward-backward recursion to compute the reliabilities of the estimated symbols. In product codes (turbo codes and serially concatenated codes) iterative decoding uses component soft-in soft-out (SISO) decoders which realize one of soft-output algorithms. The overall complexity and computational effort of the iterative decoding operation depends primarily on the complexity of the algorithms adopted in component SISO decoders. An optimal symbol-by-symbol MAP decoding method that minimizes the probability of a symbol (or bit) error is the Bahl-Cocke-Jelinek-Raviv (BCJR) algorithm [72]. The goal of this MAP decoder is to examine the received sequence and to compute the *a posteriori* probabilities of the input information bits. The BCJR algorithm was applied to turbo codes decoding in the original paper [56].

The severe drawback of the BCJR algorithm is that numerical instabilities occur if the forward and backward metrics (probabilities) are not suitably scaled at every decoding stage. The first approach to avoid this problem and, consequently, also to reduce the computational complexity of the BCJR (MAP) algorithm was proposed in [73]. In the so-called Log-MAP algorithm, the logarithms of the metrics are used. Furthermore, to avoid the sum of exponential terms, the expression known as the Jacobian logarithm is applied. The part of the Jacobian logarithm (so-called correction term) is stored in a small look-up table (LUT), as only a few values (e.g. eight) are required to achieve practically the same performance as the MAP algorithm. Hence, the complexity reduction of the Log-MAP algorithm is achieved through simple LUT accesses, instead of several calls to slow or hardware-expensive exp(x) functions.

Seeking for even more computationally efficient soft-output algorithm led to elaboration of the Max-Log-MAP algorithm [73]. It is a suboptimal derivative of the MAP algorithm, obtained from the Log-MAP by omitting the correction term. In terms of number of basic operations (i.e. additions and multiplications) for binary codes, the Max-Log-MAP algorithm is approximately twice less complex compared to the Log-MAP algorithm. This significant simplification of the Max-Log-MAP algorithm is, however, exchanged for performance degradation.

Applications of iterative decoding can also be found in the Soft-Output Viterbi Algorithm (SOVA) [74]. The SOVA is a modified Viterbi algorithm which estimates the soft output information for each transmitted binary symbol in the form of the log-likelihood function. In the opposition to the MAP algorithms, a SOVA algorithm does not require knowledge of the noise variance. The SOVA decoder can also be implemented as a sliding window decoder. In terms of decoding complexity, the SOVA algorithm requires the least amount of operations as compared with the MAP algorithms described above. For binary codes, SOVA requires about half of that of the Max-Log-MAP algorithm whereas its performance is only slightly worse than that of the Max-Log-MAP.

The algorithms mentioned above are considered the most important for product codes. However, in the literature in recent years there have appeared many modifications of them. The aim of research on new algorithms or modifications of the existing ones is to invent low-complexity algorithms, yet capable of achieving performance very close to that of the optimal MAP algorithm. Continuous need for increasing data rates in communication systems, on one hand, and advances in theory of iterative processing, on the other hand, stimulate this research.

These considerations also apply to LDPC codes. There are two basic iterative decoding algorithms for LDPC codes – both proposed in [71]: the bit-flip algorithm for iterative hard-decision decoding and iterative belief-propagation (IBP) decoding algorithm (the latter is also known as sum-product algorithm [75]). Several reduced-complexity modifications of the IBP algorithm which favourably trade off decoding complexity and error performance are presented in [76]. It is worth emphasizing that iterative decoding algorithms for turbo codes and product codes are special cases of IBP decoding [64].

An important issue closely related to decoding and to the success and deployment of iterative algorithms is quantization. Practical implementations of algorithms have strict constraints and most receiver algorithms are implemented using fixed-point arithmetic, where quantities can only take one of a finite number of values. Moreover, the passage from the optimal theoretical real-valued algorithms to their fixed-point finite precision versions is based on heuristic approaches which lead, in general, to additional suboptimality and hence, additional performance degradation. In this context, development of theoretical foundation for the design of receiver components directly in the quantized domain seems to be an interesting and well justified research topic in the area of iterative receivers [77].

3 Cognitive approaches to improved spectrum usage

As mentioned in the introduction, flexible spectrum-access methods will eventually be needed (along with others mentioned above, such as adaptivity, MIMO, advanced coding, etc.) in order to increase the efficiency of the utilization of the scarce spectrum resource. Besides the major objective of maximizing spectral efficiency, another goal of modern radio network design is to rationalize the distribution of radio resources and the cost of their usage. This means that some rationalizing mechanisms should be employed in order to balance efficiency versus fairness, high QoS versus high Quality of Experience (QoE) for both the licensed as well as unlicensed users of spectrum. Spectrally-efficient modulation techniques and associated waveforms need to be defined which would assure the opportunistic access to fragmented spectrum of dynamically-changing availability. Such techniques and waveform designs should also minimize (or at least mitigate) the interference generated to other nodes and users.

Many flexible methods have been considered for future Dynamic Spectrum Access (DSA) schemes. Many aim at spectral-efficiency maximization and are thus implemented in a centralized manner. Consequently, they necessitate relatively large overhead (feedback) traffic related to CSI in all links in the considered area. Such approaches may not be appropriate for the Cognitive Radio (CR) concept [78–80], where nodes are expected to be intelligent and autonomous entities. The cognitive decision making should be more localized and distributed and should also reflect the *rationality* paradigm, namely trading-off optimality versus cost.

3.1 Multicarrier technologies for CR

Regarding waveforms for future CR networks and systems a lot of attention has been on OFDM and related OFDMA techniques [81]. They are flexible, scalable and easy optimized waveforms, well suited for future high data-rate wireless systems and standards. It is envisioned that in the near future, when many devices will use the spectrum in an opportunistic manner, the available spectrum may be fragmented to a large number of narrow bands. OFDM has the advantage that these bands can be relatively easily aggregated to carry the Secondary User's (SU's) (i.e., an unlicensed user's) traffic, thus spectrum utilization increases significantly. Spectrum shaping can be done by assigning data traffic to available sub-carriers (SC) and nulling prohibited SCs occupied by the Primary Users (PU) (i.e., the licensed users). This non-contiguous OFDM transmission has been described in [82]. As shown in [83], various ways have been considered for limiting the interference generated by the SU to the PU. Most of them assume that this interference is dominated by the SC's neighboring the PU's spectrum. As an example, a low interference level (below −70 dB) can be achieved if a small number of SCs adjacent to a PU's spectral band are used for interference cancellation [84]. Moreover, the complexity of the FFT and IFFT algorithms in a CR non-contiguous OFDM transceiver can be lowered by removing the operations on deactivated subcarriers [82]. This operation is referred to as "pruning" and results in decreasing the FFT/IFFT execution time.

Recently, two other classes of MC transmission techniques have been considered for spectrally-efficient communication. The first one is the so-called Filtered MultiTone (FTM) modulation, based on filtering of the baseband symbols modulating multiple subcarriers. As a result, the sub-band spectra localized around the subcarriers overlap only with their closest neighbouring sub-bands, and decay rapidly to zero for more distant frequencies. The distance between the subcarriers is still equal to the inverse of the orthogonality period, and signals representing the modulating symbols are still orthogonal as in the case of OFDM. The class of Non-Orthogonal Frequency Division Multiplexing (NOFDM) was first introduced in [85], as a novel approach to MC transmission over doubly-dispersive channels affected by both time and frequency dispersion, namely both time and frequency selectivity. The underlying idea of NOMCM is filtering the modulating symbol pulses in a way that gives minimum distortion for a given Doppler spread and delay spread. As a result of the applied pulse shapes, the signals representing the data symbols are not orthogonal and overlap with signals in different symbol periods. The degree of overlapping depends on the pulse-shape also. The key advantages of NOFDM, when compared with OFDM, are: higher spectral efficiency (also because the cyclic prefix is not needed) and higher robustness against frequency-selective fading. The reception techniques for NOFDM have been studied in [86], and the application of this type of MC signalling has been studied by the European 6th Framework Program project URANUS [87].

Despite their advantages, the challenges posed by the OFDM-based or NOFDM-based CR still need to be met. Solutions should be found to specific problems such as high Peak-to-Average Power Ratio (PAPR), sensitivity to frequency offsets, and Inter-carrier Interference (ICI), as well as in conjunction to other CR challenges such as accurate and agile spectrum sensing, interference avoidance, cross-layer design and complexity, flexible RF front-end, etc.

Multicarrier technologies will probably dominate wireless communication standards in the upcoming 10–15 years. As discussed above, they are also suitable for non-standardized, opportunistic and cognitive access to wireless networks and their spectrum resources. However, there are a number of issues that will be challenges for researchers, engineers and regulatory bodies for the next 10–20 years. One is multi-band MC system design, in which the total bandwidth is divided into smaller parts. It appears that, in the CR field, the number of bands used for MC multi-band transmission depends on the total bandwidth of the spectral "holes" and their location. Other factors, such as the interference level and QoS requirements should also be taken into account. The challenges in designing a multiband MC transceiver include flexible broadband RF front-end with broadband antennas and their transmit/receive switches, complex RF circuit design, fast analog-to-digital and digital-to-analog converters of high dynamic range, wide-range frequency synthesizers, and so on. Other issues of the MC-based CR are related to the interference reduction between the users, sensing of the radio environment, synchronization of non-contiguous MC signalling with missing pilots and possibly asynchronous transmissions of multiple users, detection of the adopted transmission parameters or efficient signalling of these parameters, etc.

3.2 Distributed spectrum access and usage

Dynamic allocation of resources usually aims at the maximization of the multiuser system performance. In the case of OFDMA-based, or NOFDM-based multiple access (NOFDMA), dynamic subcarrier assignments, also called scheduling, various methods can be employed to increase the network spectral efficiency and users' perceived throughput taking some fairness measures into account. Apart from regular scheduling, adaptive bit and power allocation is also possible. For a single link, the solution of the problem of optimal power allocation maximizing channel capacity is known as the *water-filling* principle [88]. For MC transmission and multiple links, the problem is known as multiuser water-filling, and is considered quite complex in terms of finding the absolute optimum. In practical applications it can be done in two steps: first the SCs are assigned to users, and then a water-filling algorithm for each single user determines the bit and power allocation to these SCs. The adaptive subcarrier-, bit- and power-allocation methods in OFDMA have attracted considerable research interest, mainly to find some more practical algorithms for increasing spectral efficiency, fairness and the resulting QoE for an average user.

The key issue of the dynamic spectrum access and usage in the framework of CR is to assure decentralized, yet efficient and fair distribution of radio resources. This is one of the major challenges in the design of CRs, which reflects the dilemma of optimality versus complexity, efficiency versus fairness and centralized solutions with extensive control-traffic overhead versus localized sub-optimum decisions. Apart from the optimality and fairness problems, the DSA algorithms in OFDMA-based (or NOFDMA-based) CR should take other distinctive issues into account, such as dynamically changing availability of radio resources, the necessity of interference management, decision on when to use the single-band, and when the multi-band OFDM scheme. These decisions depend on various parameters: required throughput, the interference temperature, hardware limitations, computational complexity, the number of spectral holes, etc. It is expected that these issues and challenges will drive a considerable volume of research in the area of CR in the next decade.

A powerful tool to study competition and cooperation among intelligent rational decision makers is game theory. The main purpose of using it in radio resource access is to model competitive behavior and strategic interactions among network nodes (the players). The application of the economic concepts, such as competition, cooperation or the mixture thereof, allows to analyze the problem of flexible usage of limited radio resources in a competitive environment [89,90].

Game-theoretic OFDMA scheduling has been considered in the literature in two major directions: the centralized SC allocation, which allows for more efficient and fair spectrum utilization and applies cooperative game theory, Nash-Bargaining Solution (NBS); and the distributed decision making, which uses non-cooperative games and results in a Nash Equilibrium (NE) as a game solution. The motivation behind the choice of the game model reflects desired properties of the system, and various criteria of the system performance should be taken into account. The drawbacks of the existing game-theoretic solutions for the DSA problem are either lack of optimality (NBS is usually not the optimal solution), lack of rationality (the utilities are defined so as to maximize the total sum-throughput, or neglect QoE, the power economy or computational complexity), lack of generality (Pareto-optimal solutions are found for a very limited number of users) or lack of suitability for the CR networks (some optimal solutions can be found in a cooperative manner which requires centralized management). Moreover, most of these approaches assume perfect CSI, which is not the case in practice. Some promising decentralized iterative algorithms require some time for convergence to the optimum, and in most cases they require some control traffic to improve or adjust the players' strategies. Some other interesting distributed solutions for cognitive spectrum usage are based on the idea of the accessed-bandwidth limiting and processing of locally available information [91,92]. In [91], an up-link transmission is assumed, Successive Interference Cancellation (SIC) at the base-station receiver, and the fact that the users know their turn in the PIC procedure. In [92], taxation of spectral resources is considered in which the users know the tax-rate at a particular area (broadcasted by the base station), and only their own-link CSI.

Game theory has undoubtedly a great potential for bringing in the right tools to study and understand the competition of CRs for spectrum access, and for efficient spectrum usage. It can also provide useful solutions as opposed to complex optimization procedures, and therefore it can constitute one of the key research vehicles in the area of cognitive spectrum usage for the next 10–20 years.

4 Conclusions

This paper has briefly reviewed theoretical results as well as practical approaches to bandwidth- and energy-efficient radio access. The paper has identified the fundamental communication limits in fading and then discussed various technologies suitable to approaching these limits namely general coding/decoding concepts, adaptation, iterative processing, low-complexity decoding and, finally, flexible/cognitive spectrum usage. It has also identified some open issues in this broad and exciting field.

References

1. Ericsson T.: A Gaussian channel with slow fading. IEEE Trans. Information Theory **16**(3), pp. 353–355 (1970).
2. Ozarow L.H., Shamai S., Wyner A.D.: Information theoretic considerations for cellular mobile radio. IEEE Trans. Veh. Technol. **43**(2), pp. 359–378 (1994).
3. Goldsmith A.J., Varaiya P.P.: Capacity of fading channels with channel side information. IEEE Trans. Information Theory **43**(6), pp. 1986–1992 (1997).
4. Borade S., Zheng L.: Wideband fading channels with feedback. In: Proc. Allerton Conference on Communication, Control, and Computing, Monticello, IL (2004).
5. Abou-Faycal I., Trott M.D., Shamai S. (Shitz): The capacity of discrete-time memoryless Rayleigh fading channels. IEEE Trans. Information Theory **47**(3), pp. 1290–1301 (2001).
6. Gursoy M.C., Poor H.V., Verdú S.: The noncoherent Rician fading channel – Part I: Structure of the capacity-achieving input. IEEE Trans. Wireless Commun. **4**(5), pp. 2193–2206 (2005).
7. Gursoy M.C., Poor H.V., Verdú S.: The noncoherent Rician fading channel – Part II: Spectral efficiency in the low power regime. IEEE Trans. Wireless Commun. **4**(5), pp. 2207–2221 (2005).
8. Gursoy M.C., Poor H.V., Verdú S.: Spectral efficiency of peak power limited Rician block-fading channels. In: Proc. IEEE Int. Symposium on Information Theory (ISIT), Chicago, IL (2004).
9. Huang J., Meyn S.P.: Characterization and computation of optimal distributions for channel coding. IEEE Trans. Information Theory, **51**(7), p. 2336–2351 (2005).
10. Marzetta T.L., Hochwald B.M.: Capacity of a mobile multipleantenna communication link in Rayleigh flat fading. IEEE Trans. Information Theory **45**(1), pp. 139–157 (1999).
11. Zheng L., Tse D.N.C.: Communication on the Grassman manifold: A geometric approach to the noncoherent multiple-antenna channel. IEEE Trans. Information Theory **48**(2), pp. 359–383 (2002).
12. Médard M.: The effect upon channel capacity in wireless communications of perfect and imperfect knowledge of channel. IEEE Trans. Information Theory **46**(3), pp. 933–946 (2000).
13. Lapidoth A., Shamai S. (Shitz): Fading channels: How perfect need 'perfect side information' be?. IEEE Trans. Information Theory **48**(5), pp. 1118–1134 (2002).
14. Verdú S.: Spectral efficiency in the wideband regime. IEEE Trans. Information Theory **48**(6), pp. 1319–1343 (2002).
15. Lapidoth A., Moser S.M.: Capacity bounds via duality with applications to multiple-antenna systems on flat-fading channels. IEEE Trans. Information Theory **49**(10), pp. 2426–2467 (2003).
16. Gursoy M.: On the capacity and energy efficiency of training-based transmissions over fading channels. IEEE Trans. Information Theory **55**(10), pp. 4543–4567 (2009).
17. Hassibi B., Hochwald B.M.: How much training is needed in multiple-antenna wireless links. IEEE Trans. Information Theory **49**(4), pp. 951–963 (2003).
18. Shannon C.E.: Channels with side information at the transmitter. IBM J. Res. Dev. **2**, pp. 289–293 (1958).
19. Biglieri E., Proakis J., Shamai S. (Shitz): Fading channels: Information-theoretic and communications aspects. IEEE Trans. Information Theory **44**(6), pp. 2619–2692 (1998).
20. Tse D., Viswanath P.: Fundamentals of Wireless Communication. Cambridge Univ. Press, Cambridge, UK (2005).
21. Klang G. et al.: Identification of radio-link technologies. IST-2003-507581 WINNER, D2.1 Public Deliverable.

22. Steele R., Webb W.: Variable rate QAM for data transmission over Rayleigh fading channels. In: Proc. IEEE Wireless, Calgary, Canada, pp. 1–14 (1991).
23. Morinaga N., Nakagawa M., Kohno R.: New concepts and technologies for achieving highly reliable and high-capacity multimedia wireless communication systems. IEEE Commun. Mag. **35**, pp. 34–41 (1997).
24. Chua S.-G., Goldsmith A.J.: Adaptive coded modulation for fading channels. In: Proc. IEEE Int. Conf. Commun., Montreal, Canada (1997).
25. Goldsmith A.J., Chua S.-G.: Variable-rate variable-power MQAM for fading channels. IEEE Trans. Commun., **45**(10), pp. 1218–1230 (1997).
26. Chung S.T., Goldsmith A.J.: Degrees of freedom in adaptive modulation: A unified view. IEEE Trans. Commun. **49**(9), pp. 1561–1571 (2001).
27. Jafar S.A., Vishwanath S., Goldsmith A.J.: Throughput maximization with multiple codes and partial outages. In: Proc. Global Telecommun. Conf., San Antonio, TX (2001).
28. Hanzo L., Wong C.H., Yee M.S.: Adaptive wireless transceivers: Turbo-coded, turbo-equalised and space-time coded TDMA, CDMA and OFDM systems. Wiley, New York, NY (2002).
29. Keller T., Hanzo L.: Adaptive modulation techniques for duplex OFDM transmission. IEEE Trans. Veh. Technol. **49**(5), pp. 1893–1906 (2000).
30. Chow P.S, Cioffi J.M., Bingham J.A.C.: A practical discrete multitone transceiver loading algorithm for data transmission over spectrally shaped channels. IEEE Trans. Commun. **43**(2-4), pp. 773–775 (1995).
31. Fisher R.F.H., Huber B.: A new loading algorithm for discrete multitone transmission. In: Proc. Global Telecommun. Conf., London, UK (1996).
32. Rohling H., Grundheid R.: Performance of an OFDM-TDMA mobile communication system. In: Proc. Global Telecommun. Conf., London, UK (1996).
33. Münz G., Pfletschinger S., Speidel J.: An efficient waterfilling algorithm for multiple access OFDM. In: Proc. Global Telecommun. Conf., Taipei, Taiwan (2002).
34. Pfletschinger S., Münz G., Speidel J.: Efficient subcarrier allocation for multiple access in OFDM systems. In: Proc. Int. OFDM Workshop, Hamburg, Germany (2002).
35. Catreux S., Driessen P.F., Greenstein L.J.: Attainable throughput of an interference-limited multiple-input multiple-output (MIMO) cellular system. IEEE Trans. Commun., **49**(8), pp. 1307–1311 (2001).
36. Liew T.H., Hanzo L.: Space-time block coded adaptive modulation aided OFDM. In: Proc. Global Telecommun. Conf., San Antonio, TX (2001).
37. Dlugaszewski Z., Piatyszek A., Wesolowski K.: Adaptive multi-coded OFDM transmission for selective fading channels. In: Proc. IST Mobile & Wireless Communications Summit, Aveiro, Portugal (2003).
38. Wesolowski K., Dlugaszewski Z.: Information theory aspects of digital transmission over fading channels. In: Proc. National Symposium of Radio Science, Poznan, Poland (2002).
39. Hu S., Duel-Hallen A.: Combined adaptive modulation and transmitter diversity using long range prediction for flat fading mobile radio channels. In: Proc. Global Telecommun. Conf., San Antonio, TX (2001).
40. Holliday T., Goldsmith A., Glynn P.: Wireless link adaptation policies: QoS for deadline constrained traffic with imperfect channel estimates. In: Proc. Int. Commun. Conf., New York, NY (2002).
41. Falahati S., Svensson A., Sternad M., Mei H.: Adaptive Trellis-coded modulation over predicted flat fading channels. In: Proc. IEEE Veh. Technol. Conf., Orlando, FL (2003).
42. Falahati S., Svensson A., Ekman T., Sternad M.: Adaptive modulation systems for predicted wireless channels. IEEE Trans. Commun. **52**(2), pp. 307–316 (2004).
43. Koutsopoulos I., Tassiulas L.: Channel state-adaptive techniques for throughput enhancement in wireless broadband networks. In: Proc. Conference on Computer Communications, Anchorage, AK (2001).
44. Lim C.H., Cioffi J.M.: Performance of the adaptive rate MQAM with on/off power control. IEEE Communications Letters **5**, pp. 16–18 (2001).

45. Tang F., Deneire L., Engels M., Moonen M.: Adaptive link adaptation. In: Proc. Global Telecommun. Conf., San Antonio, TX (2001).
46. Tralli V., Vaccari A., Verdone R., Andrisano O.: Adaptive time and frequency resource assignment with COFDM for LMDS systems. IEEE Trans. Commun. **49**(2), pp. 235–238 (2001).
47. Ulukus S., Biglieri E., Win M.Z.: Optimum modulation and multicode formats in CDMA systems with multiuser receivers. In: Proc. Conference on Computer Communications, Anchorage, AK (2001).
48. Caire G., Shamai S.: On the capacity of some channels with channel state information. IEEE Trans. Information Theory **45**(9), pp. 2007–2019 (1999).
49. Ephremides A.: Energy concerns in wireless networks. IEEE Wireless Commun. **9**(4), pp. 48–59 (2002).
50. Minto M., Ragazzi D., Agarossi L., Giangaspero L.: Minimum transmission power algorithm for OFDM-based flexible systems. In: Proc. Workshop on Broadband Wireless Ad-Hoc Networks and Services, Sophia-Antipolis, France (2002).
51. Polydoros A., Rautio J., Razzano G., Bogucka H., Ragazzi D., Dallas P.I., Mammela A., Benedix M., Lobeira M., Agarossi L.: WIND-FLEX: Developing a novel test-bed for exploring flexible radio concepts in an indoor environment. IEEE Commun. Mag. **41**(7), pp. 116–122 (2003).
52. Dlugaszewski Z., Wesolowski K., Bogucka H., Tyczka P., Piatyszek A., Polydoros A., Nikitopoulos K., Dagres Y., Paltenghi G., Giangaspero L., Benedix M., Ertel J: Information theory results derived from WIND-FLEX implementation. 4.6 Public Deliverable of the IST-1999-10025 project WIND-FLEX (2002).
53. Wong C.Y., Cheng R.S., Ben Letaief K., Murch R.D.: Multiuser OFDM with adaptive subcarrier, bit, and power allocation. IEEE J. Select. Areas Commun. **17**, pp. 1747–1758 (1999).
54. Johansson M.: Diversity-enhanced equal access – Considerable throughput gains with 1-bit feedback, in Proc. IEEE Workshop on Signal Processing Advances in Wireless Communications, Lisboa, Portugal, July 2004.
55. Baum K.L., Kostas T.A., Sartori P.J., Clas-son B.K.: Performance characteristics of cellular systems with different link adaptation strategies. IEEE Trans. Veh. Technol. **52**(6), pp. 1497–1507 (2003).
56. Berrou C., Glavieux A., Thitimajshima P.: Near Shannon limit error correcting coding and decoding: Turbo codes. In: Proc. IEEE Int. Conf. Commun., Geneva, Switzerland (1993).
57. Wang X., Poor H.V.: Iterative (turbo) soft interference cancellation and decoding for coded CDMA. IEEE Trans. Commun. **47**(7), pp. 1046–1061 (1999).
58. Wehinger J., Mecklenbräuker C.: Iterative CDMA multiuser receiver with soft decisiondirected channel estimation. IEEE Trans. Signal Process. **54**(10), pp. 3922–3934 (2006).
59. Tuchler M., Koetter R., Singer A.C.: Turbo equalization: Principles and new results. IEEE Trans. Commun. **50**(5), pp. 754–767 (2002).
60. Kschischang F., Frey B., Loeliger H.-A.: Factor graphs and the sum-product algorithms. IEEE Trans. Information Theory **47**(2), pp. 498–519 (2001).
61. Attias H.: A variational Bayesian framework for graphical models. Advances in Neural Information Processing Systems, pp. 209–215 (2000).
62. Beal M.: Variational algorithms for approximate inference. Ph.D. dissertation, University of Cambridge (2003).
63. Cover T.M., Thomas J.A.: Elements of Information Theory, 2nd ed. Wiley Interscience, New York, NY (2006).
64. McEliece R., MacKay D., Cheng J.: Turbo decoding as an instance of Pearl's belief 'propagation' algorithms. IEEE J. Select. Areas Commun. **16**(2), pp. 140–152 (1998).
65. Worthen A.P., Stark W.E.: Unified design of iterative receivers using factor graphs. IEEE Trans. Information Theory **47**(2), pp. 843–849 (2001).
66. Minka T.: Divergence measures and message passing. Microsoft Research Ltd., Tech. Rep. (2005).
67. Boutros J., Caire G.: Iterative multiuser joint decoding: Unified framework and asymptotic analysis. IEEE Trans. Information Theory **48**(7), pp. 1772–1793 (2002).

68. Novak C., Matz G., Hlawatsch F.: A factor graph approach to joint iterative data detection and channel estimation in pilot-assisted idma transmissions. In: Proc. IEEE Int. Conf. Acoust., Speech, Signal Process, Las Vegas, NV (2008).
69. Stefanatos S., Katsaggelos A.: Joint data detection and channel tracking for OFDM systems with phase noise. IEEE Trans. Signal Process. **56**(9), pp. 4230–4243 (2008).
70. Hu B., Land I., Rasmussen L., Piton R., Fleury B.: A divergence minimization approach to joint multiuser decoding for coded CDMA. IEEE J. Select. Areas Commun. **26**(3), pp. 432–445 (2008).
71. Gallager R.G.: Low density parity check codes. IRE Trans. Inform. Theory **8**, pp. 21–28 (1962).
72. Bahl L., Clocke J., Jelinek F., Raviv J.: Optimal decoding of linear codes for minimizing symbol error rate. IEEE Trans. Information Theory **20**(3), pp. 284–287 (1974).
73. Robertson P., Villebrun E., Hoeher P.: A comparison of optimal and sub-optimal MAP decoding algorithms operating in the log domain. In: Proc. IEEE Int. Conf. Commun., Seattle, WA (1995).
74. Hagenauer J., Hoeher P.: A Viterbi algorithm with soft-decision outputs and its applications. In: Proc. IEEE Global Telecommun. Conf., Dallas, TX (1989).
75. Wiberg N.: Codes and decoding on general graphs. Ph.D. dissertation, Dept. of Electrical Engineering, Linköping, Sweden, 1996, Linköping Studies in Science and Technology.
76. Fossorier M.P.C., Mihaljevic M., Imai H.: Reduced complexity iterative decoding of low-density parity check codes based on belief propagation. IEEE Trans. Commun. **47**(5), pp. 673–680 (1999).
77. NEWCOM++ WPR.4, DR.4.1: State of the art. Tech. Rep. (2008).
78. Mitola J.: Cognitive radio: An integrated agent architecture for software defined radio. Ph.D. dissertation, Royal Institute of Technology (KTH), Sweden (2000).
79. Mitola J.: Cognitive radio architecture: The engineering foundations of radio XML. Wiley & Sons, Hoboken, NJ (2006).
80. Haykin S.: Cognitive radio: Brain-empowered wireless communications. IEEE J. Select. Areas Commun. **23**(2), pp. 201–220 (2005).
81. Hanzo L., Munster M., Choi B.J., Keller T.: OFDM and MC-CDMA for broadband multi-user communications, WLANs and broadcasting. Wiley & Sons, Hoboken, NJ (2003).
82. Rajbanshi R., Wyglinski A.M., Minden G.J.: OFDM-based cognitive radios for dynamic spectrum access networks. In: Hossain E., Bhargava V.K. (Eds.): Cognitive Wireless Communication Networks. New York, NY: Springer Science and Business Media (2007).
83. Mahmoud H.A., Yucek T., Arslan H.: OFDM for cognitive radio: Merits and challenges. IEEE Wireless Communications **16**(2), pp. 6–14 (2009).
84. Brandes S., Cosovic I., Schnell M.: Reduction of out-of-band radiation in OFDM systems by insertion of cancellation carriers. IEEE Commun. Letters **10**(6), pp. 420–422 (2006).
85. Kozek W., Molisch A.F.: Nonorthogonal pulseshapes for multicarrier communications in doubly dispersive channels. IEEE J. Select. Areas Commun. **16**(8), pp. 1579–1589 (1998).
86. Hunziker T., Dahlhaus D.: Iterative symbol detection for bandwidth efficient nonorthogonal multicarrier transmission. In: Proc. IEEE Veh. Technol. Conf., Tokyo, Japan (2000).
87. IST-027960 URANUS, D3.1: GMCR transceiver for air interfaces in single mode operation. Tech. Rep. (2007).
88. Goldsmith A.: Wireless Communications. Cambdrige, UK: Cambridge Univ. Press (2005).
89. MacKenzie A.B., DaSilva L.A.: Game Theory for Wireless Engineers. Morgan & Claypool, San Rafael, CA (2006).
90. Srivastava V., Neel J., MacKenzie A.B., Menon R., DaSilva L.A., Hicks J.E., Reed J.H., Gilles R.P.: Using game theory to analyze wireless ad hoc networks. IEEE Commun. Surveys & Tutorials, **7**(4), pp. 46–56 (2005).
91. Perlaza S.M., Debbah M., Lasaulce S., Bogucka H.: On the benefits of bandwidth limiting in decentralized vector multiple access channels. In: Proc. Int. Conf. Cognitive Radio Oriented Wireless Networks and Communications, Hannover, Germany (2009).
92. Bogucka H.: Linear and arithmetically-progressive taxation of spectrum resources in cognitive OFDMA. In: Proc. Int. Workshop on Game Theory in Communication Networks, Pisa, Italy (2009).

Multimedia signal processing for wireless delivery

Pierre Duhamel (✉), Enrico Magli, Gabriella Olmo, Christine Guillemot, Michel Kieffer, and Claudio Weidmann

1 Wireless multimedia delivery: The situation

Clearly, when one searches to build communication systems to mobiles or between mobile terminals with a very high capacity, he implicitly has in mind that some multimedia (high quality sound, still images, video, and more in the near future) has to be transmitted. Speech by itself would not justify such investments. Even when web surfing is considered, the files by themselves are not very demanding, only the subparts containing images and multimedia are costly in terms of bit-rate.

Therefore, it has been recognized that it is useful to tune the communication system globally, taking this fact into account. This approach leads naturally to what is known as *cross layer optimization* (see [1] for a general approach of cross layer optimization and [2] for an application to video streaming) and *joint source and channel coding/decoding* (JSCC/D) approach [3,4].

The difficulty is that the layered architecture was designed in order to assign a specific task to each layer, the layers assuming that the previous one was successful in completing its task. Therefore, they are independent of each other, and, as a result, the delivery is independent of the type of signal. This universality of the network should not be put into question, even if some global optimization is performed.

The techniques cited above are relatively short term. While cross layer optimization is already used, JSC Decoding should be usable soon, due to recent advances, and JSC Coding requires a further global optimization, which may take some time [4].

However, the evolution of wireless multimedia will certainly go beyond the mere transmission of speech, sound, images, videos and html files, and such an expansion will require further scientific investment in a number of fields. This will be driven by applicative needs, therefore we first spend some time on examining the corresponding trends.

Benedetto S., Correia L.M., Luise M. (eds.): The Newcom++ Vision Book. Perspectives of Research on Wireless Communications in Europe.
DOI 10.1007/978-88-470-1983-6_7,

1.1 General trend

The studies on multimedia transmission were sometimes motivated by somewhat erroneous motivations. Initially, picture phones were studied more than 30 years ago, based on analogue phones. Then, when digital compression of video became mature, a first standard targeting visiophony was proposed (H.263) [5], and the engineers thought that it could be a widely used application. This turned out to be false. The same opinion arose about 10 years ago, with the arrival of mobile phones, and studies on their use were predicting a large success to picture communication. Recently, it was again claimed that, since new phones now have cameras, the time for visiophony has come. This does not imply that the emergence of visiophony will not happen this time, even if one can reasonably have some doubts.

Conversely, who would have predicted the success of SMS, which are so rustic, but are now so widely used? Moreover, given the fact that SMS were popular, who would have guessed the relative lack of use of the MMS, even for mobile phones incorporating a camera?

In the recent times, some usage of mobile phones increased suddenly, such as video consultation or streaming (clips, youtube and related).

On January 1st, 2006, France had more than two million subscribers to "broadband" mobile services. Video was the first application used by these customers with approximately 60% of users watching television programs (fifty channels), the rest of users watching downloaded videos (video on demand – VOD of news, music videos or charm).

In VOD denoted as Mobile TV, Orange counted 22.3 million sessions viewing in 2005, including 4.5 million just for the month of December. The length of a video session was 40 minutes per month. On his side, SFR delivered 4.3 million TV-video sessions, and 1.2 million in December. Since then, the demand in terms of video kept increasing, with a shift in usage: the demand for broadcasted programs is increasing less than the VOD, except on specific occasions (such as sport events), where the number of viewers is huge.

Combined with the large increase in web browsing, video downloading (either streaming, VoD, or podcasts) seems to be the current application that drives the increasing demand in capacity of mobile systems.

Obviously, when one observes what happened in these applications, it should be clear that the fact that an application involves a more advanced technology is not a criterion for its wide success (see visiophony), and certainly no more than its adhering to some sociological trends. Therefore, one should project in the mobile context the trends that are already observed in the wireline setting, and guess which adds-on could be useful.

1.2 The driving forces

It is likely that mobile Internet will be a driving context for a while, and that applications already popular on fixed terminals will also have some success on mobile terminals.

Besides multimedia downloading ("hidden" or not behind web browsing, consultation of private media centres or video surveillance), games and peer to peer communications seem to be the first candidates for a successful declination on mobile devices. Applications linked to localization will also likely be successful.

Multimedia will also take its full meaning, even if downloading only is considered. The first screens for 3-D TV are ready; some of them are even usable without specific glasses. In the longer timeframe of this report, it is possible that these screens will be widely diffused for classical TV, and available for mobile devices. Furthermore, stereo-vision based images [6] should not be confused with 3-D object representation [7], which allow seeing one object from any possible point of view, even on classical screens, by rotating them.

This corresponds to a large increase of the required bandwidth for providing all users with the corresponding data. In fact, even if stereo vision requires a moderate amount of additional data (from 10 to 20%), they will certainly be used with good quality video, and true 3D data are large intrinsically. Such an increase may saturate conventional networks. Therefore, due to scarcity of mobile resources, the point to multipoint situation (e.g. broadcasting) should be used when possible, to save bandwidth. This trend to Multiple-transmitter, multiple-receiver situation is already perceived in the scientific publications, and is present in the standards (even if not fully implemented yet), but will increase and will be extended to heterogeneous situations, including peer to peer communications. It has to be noticed also that the broadcasting/multicasting techniques that are standardized could be largely improved by more sophisticated techniques involving more cross layer.

Beyond downloading, games, and peer to peer communications, it may happen that other multimedia services could have some specific declination on mobile networks. Some of them are outlined later in this paper.

1.3 General consequences on the time frame

It appears that some of the possible applications listed above depend on the availability of devices or systems that are on their way: screens for stereo vision, MANETS for peer to peer communication [8]. When these devices or systems will have sufficient distribution, their use in the context of wireless multimedia can develop very quickly.

Other key factors are knowledge-related. Some of the items listed below need the cooperation of a variety of knowledge often found in different fields, and even exploited in terms of implementation by different people. A typical example is the Joint Source and Channel Decoding (JSCD, see below). Designing a receiver that can take the best of the received signals (involving some channel related quantities, source bitstream, as well as network generated parts) requires a variety of knowledge. Moreover, assuming that simulations have proven that Joint Source Channel and Network Coding (JSCNC) is an efficient alternative to classical receivers, the next difficulty is that the design of the various parts of the classical receivers (channel, network, source) are performed by separate teams, while the design of a "new" receiver would also require a mixture of their skills.

This is the reason why some topics that have been already studied for a while are not implemented in practical systems, even if they prove to be efficient. In these cases, more work is necessary to convince a variety of people.

Especially in such a context, two complementary approaches are necessary: improvements of basic techniques and technologies on one side, and a global optimization on another side, optimization which is linked to the fact that a large part of the throughput on the wireless link has specific properties and constraints introduced by various parts of the network.

2 Enabling technologies

Obviously, the evolution of wireless multimedia transmission and processing is driven by advances in devices, in the networks, and in the processing. Some of these enabling technologies which are likely to have an impact in a foreseeable future are listed below.

2.1 3-D

The fact that in a near future broadcasted images could be 3-D may seem unrealistic, but things are moving very fast, since the technology seems to be close to mature. A first example is given below.

Cable TV providers will begin 3D TV broadcasting test for the first time in Korea in March 2010 at the earliest. This is thought of as the next TV revolution that will follow black and white, colour and HD TV. According to the government and related industry, a major IT services provider (KCC) was confirmed to discuss test broadcasting with fee-based broadcasters such as cable TV in order to provide 3D TV to the public. If KCC and cable TV continue to cooperate as expected, MSO will begin test service in March in metropolitan area.

Test broadcasting needs polarizing glasses to watch TV but KCC is considering 3D screen broadcasting according to recent technology development. An insider of KCC said, "We are considering starting 3D TV test broadcasting first to digital cable TV that has enough frequencies."

Obviously, the key for transposing this advance to wireless broadcasting is source compression, which will have to be very efficient for allowing such a bandwidth on mobile channels. However, this target now seems realistic in a medium time frame.

2.1.1 Stereo vision

The term says all: just like stereo sound, where two loudspeakers allow locating the direction of sounds, here, two (or more) visual presentations recreate the third dimension. This is usually denoted as *3D Film.*

The term 3D or 3-D is used to describe any visual presentation that attempts at creating an illusion of depth to the viewer. Generally this is accomplished by filming two images at the same time from different angles. The images are then displayed in such a way that the viewers visual cortex will interpret the pair of images as a single 3D image. There are several techniques to produce such 3D films. They all involve projecting stereoscopic images:

- anaglyphic 3D film (with glasses);
- polarization 3D film (with glasses);
- alternate-frame sequencing (with glasses);
- autostereoscopic displays (without glasses).

Just like the audio case, if such a 3-D film has to be transmitted, source compression methods must be adapted. This topic has not been widely studied, and more effort must be put especially on stereo vision models in order to better understand the interaction between the several views. It should be remembered that the binaural psychoacoustic model was the key for bit rate reduction of high quality audio.

2.1.2 3-D signal processing

3D Film (based on stereovision, as explained above) and the technique behind 3D film is not to be mistaken to be anything like the technique behind the term '3D Games'. Today, generally, 3D Games refer to a game that is set in a virtual 3D world where you can move about in all directions. But the eye (or brain) is not really 'seeing' the images as 3D, not more than viewing a 2D photograph of our 3D world. But with most 3D film techniques available, 3D Games could easily be made to generate realistic moving 3D images.

Moreover, the techniques used to represent 3-D objects in these virtual 3-D worlds could generate more applications in connection with mobile receivers.

In fact, 3D objects (either static or moving) play an increasing role in many industrial areas, such as medical imaging, computer aided design, games, and more generally "digital leisure".

These objects are usually represented by polygonal meshes in 3D or 3D+t, which correspond to a huge amount of data because of the require accuracy and the amount of carried information. Therefore, generation of transmission of these representations requires 3D (static and moving) mesh compression, segmentation, multi-resolution analysis, simplification of existing meshes, and many other tools.

A first use of these "true" 3-D objects is *augmented reality*. Augmented reality (AR) [9] is a term for a live direct or indirect view of a physical real-world environment whose elements are merged with (or *augmented* by) virtual computer-generated imagery – creating a mixed reality. The augmentation is conventionally in real-time and in semantic context with environmental elements, such as sports scores on TV during a match. With the help of advanced AR technology (e.g., adding computer vision and object recognition) the information about the surrounding real world of the user becomes interactive and digitally usable. Artificial information about the environment and the objects in it can be stored and retrieved as an information layer on top of the real world view. Commonly known examples of AR are the yellow "first down" line seen in television broadcasts of American football games, and the coloured trail showing location and direction of the puck in TV broadcasts of hockey games. The real-world elements are the football field and players, and the virtual element is the yellow line, which is drawn over the image by computers in real time. Similarly, rugby fields and cricket pitches are branded by their sponsors using Augmented Reality; giant logos are inserted onto the fields when viewed on television. Television telecasts of swimming events also often have a virtual line which indicates the position of the current world record holder at that timing.

Another type of AR application uses projectors and screens to insert objects into the real environment, enhancing museum exhibitions for example. The difference to a simple TV screen for example, is that these objects are related to the environment of the screen or display, and that they often are interactive as well.

The difference between stereo and 3-D can be understood easily from the following example: while stereo provides an illusion of depth, a 3-D processing would give the viewer the ability to watch World Cup best moments through the eyes of every player on the pitch or from any spot in the stadium in 3D mode.

2.2 Screens and cameras

All these applications are made feasible by recent advances in the technology of cameras and screens. A manufacturer has recently shown off a new single-lens camera able to capture 3D images.

The majority of existing 3D set-ups use two-camera systems to record images tailored specifically for the left and right eye of the viewer. The new camera takes a single image that is split by mirrors and recorded on two sensors, resulting in a "smoother" picture, according to the manufacturer.

Concerning screens, several companies proposed screens which do not need the use of special glasses.

These televisions use the familiar trick of sending slightly different images to the left and right eyes – mimicking our stereoscopic view of the real world. But where old-fashioned 3-D movies rely on the special glasses to block images meant for the other eye, the new technology places tiny lenses over each of the millions of red, green and blue sub pixels that make up an LCD or plasma screen. The lenses cause each sub pixel to project light at one of nine angles fanning out in front of the display.

A processor in the TV generates nine slightly different views corresponding to the different angles. From almost any location, a viewer catches a different image in each eye.

Providing so many views is key to the dramatic results. Displays that project just two views require the audience to sit perfectly still in front of the screen, while with the recent technology, viewers can move around without losing much of the effect – one set of left/right views slips into another, with just a slight double-vision effect in the transitions.

2.3 The network

2.3.1 Mobile ad hoc networks (MANETs)

A mobile ad hoc network (MANET), sometimes called a mobile mesh network, is a self-configuring network of mobile devices connected by wireless links.

Each device in a MANET is free to move independently in any direction, and will therefore change its links to other devices frequently. Each must forward traffic unrelated to its own use, and therefore be a router. The primary challenge in building a MANET is equipping each device to continuously maintain the information required to properly route traffic.

As such, MANETs are addressed in the paper by J. Pérez-Romero et al. (the second paper of this volume). They are cited here as an enabling technology which can support several applications involving multimedia, and are sometimes even a necessity for some situations to develop.

The first, obvious one is peer to peer communication and streaming, which can benefit from the wireless context (mobility is useful in this framework). However, the capacity of such networks can be largely impacted by multimedia transmission, and a careful optimization taking into account the type of signal that is circulating is needed. As a result, the global optimization described in Section 3 will be very important in this context.

2.3.2 Network coding

This technique was recently proposed for use in wired networks, and is now actively studied in the context of wireless networks. Although already mentioned in the paper by C. Antón-Haro et al. (the fourth paper of this volume), we spend some time on its description because it is one of the key technologies that can change a lot the way one is doing wireless multimedia transmission. It is also a somewhat detailed example of the claims that a global optimization is required when transmitting multimedia over wireless channels.

The concept. A traditional network makes use of routers to direct and transfer messages. The elementary function of a router is to copy each input message and forward it, as it is, to the appropriate output. The network coding (NC) paradigm brings the main novelty of allowing "processing" of messages at each intermediate hop in the network. In particular, intermediate nodes are allowed to mix (compute a certain function in some algebraic domain) incoming messages, e.g., a symbol-by-symbol linear combination, and to forward the combined packets towards the destination nodes. The encoding ensures that any destination node can receive with high probability enough combinations to recover the original messages. Thus, NC extends the concept of "encoding" of a message beyond source coding (for compression) and channel coding (for protection against errors and losses), introducing a coding stage at the network layer. This stage has been shown to increase network throughput compared to traditional networks, achieving network capacity for multicasting from a single source. Moreover, NC can also reduce delay and energy consumption, provide robustness to network dynamics, increase security, and simplify routing.

The existing NC techniques have been developed in the context of information theoretic research. Therefore, they are mostly based on ideal data and network models, such as error-free links, no delays, and synchronised nodes. A challenge is to consider NC in a multimedia transmission context, and to develop the global theory needed to fully exploit its potential. This requires addressing several challenges specific to video communications, such as controlling the quality of service (QoS), controlling the delay, optimising scheduling of packets, and jointly designing source, channel and network codes. The NC-based video communication paradigm is expected to provide significant advantages in many communication scenarios, especially video broadcast and multicast. Wireless video communications and peer-to-peer (P2P) live video streaming are only two examples of multimedia applications that would greatly benefit from NC; others include communications in wireless mesh networks, data gathering and forwarding in sensor networks, network analysis and network security, just to mention a few.

NC theory is now mainly well understood, and its huge potential has been proved in ideal scenarios. However, a lot of work is needed to put it in a practical application environment. The recent technological developments in the field of wireline and wireless multimedia communications have shown that source and channel coding need to be tailored to specific data and application environments. For video coding, a key driver has been the integration among compression, channel coding and networking, i.e., the development of network-friendly compression and error-resilience techniques, such as unequal error protection, scalable video coding and multiple description coding. Nevertheless, delivering real-time video over wireless networks is still a challenge due to scarce bandwidth, noisy links, and low power consumption requirements.

A new architecture for real-time video communication and delivery that leverages on NC could achieve high end-to-end video quality. This could lead to the following results:

- to achieve increased throughput for video communication using NC. This requires developing NC techniques tailored to video communications, which need to be network-friendly, capable of handling continuous data streams, and have low complexity to allow real-time processing and low power consumption for handheld devices;
- to provide QoS for video using NC. This requires developing application-layer QoS techniques that are designed for an NC-based transmission system. Such system will react differently than a classical network to delays, packet losses, congestion, node failures and other impairments. Therefore, the techniques used to cope with these problems must be investigated and designed;
- to demonstrate the improvements of network-coded video in practical applications.

NC: Going beyond the state-of-the-art. According to a recent technology forecast by Cisco Systems, Internet television is the application that will see the most increase in traffic in the next years. While video streaming is a consolidated technology, it is not by any means a mature field. Currently, YouTube allows smooth streaming of video sequences at 320×240 resolution with 15 frames per second. On the other hand, high-definition television applications require at least 1280×720 resolution with 30 frames per second. This difference between bandwidth available and needed for high-quality video streaming is enormous and calls for a paradigm shift. Many content providers are considering P2P networking as a new platform over which video contents can be delivered. However, this technology alone cannot solve all problems; while it indeed takes a lot of burden off the server, it introduces new communication problems, such as delay, latency, security/trust issues, and node churn. NC is likely to be an enabling technology for high-quality live video streaming. We focus on two very challenging and relevant scenarios: wireless video and P2P video streaming. The throughput increase that NC promises can significantly alleviate the bandwidth issue. In a wireless environment, it can decrease energy consumption by using less transmission attempts to communicate a given amount of information. In a P2P environment, it can reduce delay, increase security by using NC as hashing method, simplify the overlay management, and provide robustness to node failures.

As has been said, although NC has an extremely high potential, the research carried out so far is mainly of theoretical nature. Demonstrators have been developed for a few application scenarios, e.g., improving throughput in wireless networks by inserting a coding stage between the IP and MAC layers. The major industrial players in the video communication field are investigating NC for non real-time applications such as file distribution. However, it should be noted that the extension of the existing NC systems to real-time video communications is far from trivial. Delay-constrained applications call for the adoption of QoS control mechanisms to achieve the desired degree of quality. Moreover, video packets exhibit characteristics such as unequal importance, delays and playback deadlines, which require changing the way NC is performed, and the way applications use it. Therefore, it is believed that both foundational research and R&D are needed to exploit the full potential of NC for video applications.

In a future perspective, NC will act as intermediate coding layer allowing to increase throughput, and reduce delay and energy consumption. It will be coupled with application-layer QoS strategies in order to define a full-featured next-generation video coding architecture. It is expected that video communication systems based on this architecture will solve a number of problems that are currently limiting the deployment of high-quality video systems. In particular, such systems are expected to provide higher throughput, which, in conjunction with suitable QoS policies, will enable the transmission of higher quality video, with increased robustness to packet losses or failures of intermediate nodes, reduced delay/latency, and reduced energy consumption crucial for wireless applications. Applications are foreseen in many video systems, and particularly in broadcast and multicast scenarios, e.g., live television and personal video broadcast, using wireless and wireline communication systems such as 3G, WiMax, DVB, DMB, DTMB, and the Internet.

Network coding algorithms for video. It is necessary to develop NC algorithms tailored to video streaming applications. The algorithms will provide the following set of functionalities.

- Unequal error protection. The algorithms shall provide differentiated packet treatment, enabling a more reliable delivery of the most important video packets, through their identification and involvement into more coding operations. Through unequal error protection, NC will provide smooth video quality.

- Multiple description coding. Existing NC algorithms are expected to provide very poor video quality if few packets are received. Decoding for video applications should provide incremental quality as more packets are received, e.g., applying multiple description coding techniques to NC.
- Fully distributed versus centralised network coding. Fully distributed NC algorithms are extremely simple, but suboptimal. A centralised network code optimises the end-user probability of correct decoding based on full or partial knowledge of the packets they have already received. Partially distributed algorithms will also be considered in order to limit the communication overhead.
- Low-complexity decoding. Although it has been shown that NC complexity is manageable, especially the wireless applications could benefit from simpler algorithms. Therefore, it is foreseen to develop such algorithms using techniques borrowed from sparse random coding theory and digital fountain codes.

Application-layer QoS and error control techniques for network coding. Research in this field can be expected to focus on the following aspects:

- *Forward error correction (FEC) coding.* As a NC environment behaves similarly to a broadcast channel, even for point-to-point communications, it is not clear whether retransmission or FEC codes, or a hybrid thereof, will provide the best performance, and which FEC strategy should be used.
- *Joint source/network coding.* All elements of the video compression chain need to be rethought in light of the NC communication environment. This includes the design of scalable video coding and/or multiple description coding schemes, optimised for NC, which can provide the desired QoS.
- *Packet scheduling.* Another important element of QoS control is packet scheduling, as the selection of which packets must be combined and forwarded is expected to have a significant impact on the end-to-end performance. Packet scheduling algorithms tailored to NC should be designed taking into account rate-distortion information and deadlines of video packets.

3 The global optimization for wireless multimedia

In many situations, when several mobile users are asking for a given media at the same time, they will not have the same link capacity, and some of them can have a very small one. Since the delay must remain small, retransmissions must remain limited (if not forbidden). Moreover, each terminal may need a different signal quality. This is the motivation for the use of scalable signals (the bitstream contains the various qualities needed) and multiple descriptions (the more descriptions you get, the better the signal quality) on the source side. This has a number of consequences on the whole communication system, since all layers must be adapted to this point to multipoint situation.

Moreover, there is much to be gained in terms of bandwidth savings and performance improvement when considering a multicast situation: a number of registered users are requiring the same signal, may be with different qualities. Most known techniques for robust transmission of multimedia have to be adapted to this situation. Layered coding is known to be the theoretical solution from an information theory point of view, but a lot of work has to be undertaken in order to be able to build a full, practical system making optimal use of this property all along the protocol layers.

3.1 Cross layer optimization

For successful communication of multimedia data, and especially video, the network must behave as if it is predictable and ss. But wireless networks are not known for predictability and stability, but rather variability and scarcity of resources. Oversizing, the usual way of coping with variability, is not practical in wireless systems due to the scarcity of resources. Yet, the number of applications for video broadcasting over wireless channels keeps increasing. Therefore, there is a growing need for techniques that provide sufficient received video quality without overspending the scarce resources of the wireless network. However, backward compatibility is also a must, since many of these applications are already launched.

In the classical approach to network design, the networking task is partitioned into distinct layers, where each layer, assuming that the lower layers behave perfectly, attempts to provide perfect information to the upper layers. This is not fully compatible with a full optimization of the resources and this is why cross-layer communication protocols have become of great interest to some who, motivated by multimedia applications, have worked on jointly adapting the characteristics of each layer to the type of signal that was transmitted. This cross layer optimization is a first step towards a global optimization, but has the strong advantage of being backward compatible, since it does not require many changes in the protocol itself. It is expected that this approach will be more and more studied in the near future.

3.2 Joint source-channel-network-protocol coding [3]

In recent years, so-called "joint" approaches have been proposed for improved performance and better use of resources. In joint approaches, the network layers become more transparent, so that the information previously available at only one layer is now accessible to other layers. The downside to this ambitious approach is a loss of the architectural coherence that was the primary driving force behind the use of decoupled layers.

Motivated by the limits of classical approaches and by the practical challenges raised by cross-layer techniques, an interest in Joint Source/Channel Decoding (JSCD) emerged, whereby the goal is to extract the best performance out of the received signal without considering changes to the way this signal is transmitted. Clearly, the ability to use JSCD with existing standards makes it potentially very practical. Therefore, there is a good possibility that JSCD tools could become useful in the near future, provided that all of the different network layers can be incorporated in the process.

However, the first proposals in this direction were somewhat ideal compared to practical situations: in particular, most network layers were considered as fully transparent, and only the physical and application layers were considered. This approach is now changing quickly, and many works now address the whole protocol stack in a "joint" manner. In some sense, this is now a "cross layer" approach of the "joint source and channel" decoding. In this way, JSCD tools become practical when network layers are not considered as fully transparent and dedicated to a single task, and, on the other hand, the cross-layer approach benefits from additional efficient tools.

This approach will have to be extended to take into account any redundancy introduced by the network, and more specifically Network Coding.

Finally, this approach can be understood as making the best use of any redundancy found in the bitstream, either left by the source encoder, or introduced by the protocol (headers, CRCs, checksums), the network (Network Coding), and the physical layer (channel coding, modulation). However, in this JSCD approach, even when protocol layers are involved, it is easily seen that one level of optimization is not used : due to backward compatibility, the allocation of the redundancy along the whole system is not performed. This is the next step in this direction, which would result in a true joint source/protocol/network/channel optimization.

Finally, one should realize that, even if the research has concentrated mostly on video, other types of multimedia signals should be considered, such as audio, html files, and even Stereo TV and 3-D objects, as argued previously.

3.3 Services and usage: Wireless multimedia processing, beyond "plain" delivery

Finally, the global trend can be seen to be a more global optimization of the wireless network, including properties of the source into account. This should be done ideally without losing the universality of the network (which should still be able to carry various applications and signals).

The tendency towards a variety of application and service (more than plain transmission) can also be seen in the recent years. Therefore, it seems that on top of the integration of the various aspects: source, channel, protocol, network, the service should also be integrated (more than just the source). In this sense, one will then be able to have true multimedia available, rather than simple video transmission:

Such applications could be augmented reality (see above), machine to machine communication, content retrieval, and many situations that are currently studied in the multimedia setting.

References

1. Lin X., Shroff N.B., Srikant R.: A tutorial on cross layer optimization in wireless networks. IEEE J. Select. Areas Commun. **24**(8), pp. 1452–1463 (2006).
2. Choi L.U., Kellerer W., Steinbach E.: Cross layer optimization for wireless multi-user video streaming. In: Int. Conf. Image Processing, Singapore, Singapore (2004).

3. Duhamel P., Kieffer M.: Joint source-channel decoding. A cross-layer perspective with applications in video broadcasting over mobile and wireless networks. Academic Press, Oxford, UK (2010).
4. Azami S.Z., Duhamel P., Rioul O.: Joint source-channel coding: Panorama of methods. In: Proc. CNES Workshop on Data Compression, Toulouse, France (1996).
5. ITU-T Recommendation H.263, Video coding for low bit rate communication (2005).
6. Merkle P., Brust H., Dix K., Muller K., Wiegand T.: Stereo video compression for mobile 3D services. In: Proc. 3DTV Conference, Potsdam, Germany (2009).
7. Dugelay J.-L., Baskurt A., Daoudi M. (Eds.): 3D Object Processing: Compression, Indexing and Watermarking. Wiley & Sons, West Sussex, UK (2008).
8. Yan L.: Can P2P benefit from MANET? Performance evaluation from users' perspective. In: Proc. Int. Conf. Mobile Sensor Networks (2005).
9. Azuma R., Baillot Y., Behringer R., Feiner S., Julier S., MacIntyre B.: Recent advances in augmented reality. IEEE Computer Graphics **21**(6), pp. 34–47 (2001).
10. Katti S., Rahul H., Hu W., Katabi D., Medard M., Crowcroft J.: XORs in the air: Practical wireless network coding. IEEE/ACM Trans. Networking **16**(3), pp. 497–510 (2008).

Enabling "hard" technologies for future wireless

Dominique Noguet (✉), Marc Belleville, Venkatesh Ramakrishnan, Guido Masera, Dominique Morche, Chistophe Moy, and Gerd Asheid

1 Trends in silicon technologies

Wireless technology has been benefiting from the advances the silicon technology has offered in the 90s and 00s. The whole telecommunication mutation from the analog domain to the digital realm has in fact been made possible by the miniaturisation of transistors, leading to higher density and complexity though with low power efficient ICs. In turn, this move to digital communication has created the boom of the digital ICT at all layers, from broadband communications to multimedia services. With this in mind, it would not make sense to foresee what telecommunication will offer in the future without considering the trends in silicon technology research and industry.

The key factor behind the digital revolution has been the CMOS technology down-scaling following the so-called Moore's Law [1] which coined in 1965 that the number of transistors that can be placed inexpensively on an integrated circuit doubles approximately every two years. Although this rule has been validated over the 40 last years by the silicon industry and by the ITRS [3], it is agreed that we are coming to a new era where this rule will be difficult to maintain. Several reasons can be identified:

- the physics of silicon introduces side effects in nanometre CMOS technologies;
- predictability of transistors behaviour is getting less accurate, leading to lower performance, lower yield or less optimal usage of silicon;
- power density is going to levels beyond what cooling can offer;
- static power increases which makes the assumption that the overall power consumption decreases significantly as transistors shrink no longer valid;
- investments needed for new deep submicron CMOS are being so huge that only less than a handful of application justifies it.

For all these reasons, it is likely that we are on the verge of significant changes in the silicon capability roadmap, which make the analysis of future trends useful. Indeed, it is foreseen that the roadmap has to move from a pure down-scaling to

Benedetto S., Correia L.M., Luise M. (eds.): The Newcom++ Vision Book. Perspectives of Research on Wireless Communications in Europe.
DOI 10.1007/978-88-470-1983-6_8, © Springer-Verlag Italia 2012

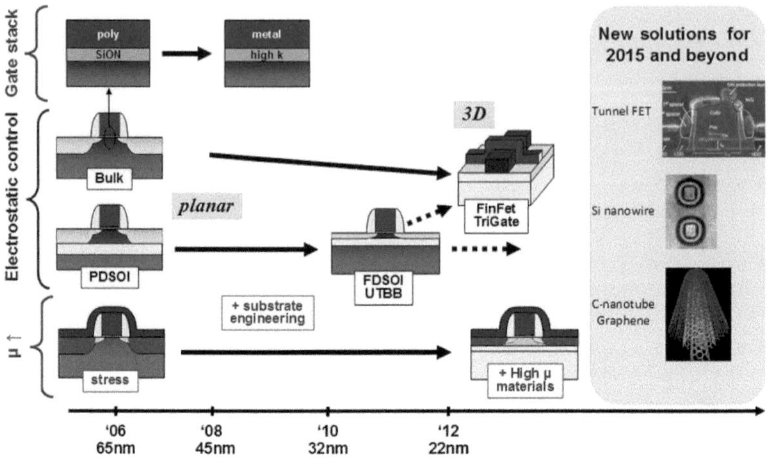

Fig. 1. Technology nodes of CMOS down-scaling with possible alternatives for post-2015 nanotechnology

new functionalities and combined technology-system innovation in order to manage future power, variability and complexity issues. However, there is no accepted candidate today to replace CMOS devices in terms of the four essential metrics needed for successful applications: dimension (scalability), switching speed, energy consumption and throughput [6,11]. Moreover, when other metrics such as reliability, designability, and mixed-signal capability are added, the dominance of CMOS is even more obvious. It is then realistic to think that other micro or nano-technologies should be seen as future add-ons to CMOS and not as a substitute for it [16]. This transition between the "business as usual" era and the entry to the post 2015[1] period where new alternative or complementary solutions need to be found is depicted in Fig. 1.

Bearing in mind this disruptive future and rather than extending the technology evaluation proposed by the ITRS, technology analysis carried out in Europe by Medea experts [16] suggests considering three major paradigms (Fig. 2):

- More Moore: corresponding to ultimate CMOS scaling;
- More than Moore: corresponding to the use of heterogeneous technologies such as MEMs or MOEMs;
- Beyond CMOS: corresponding to nanotechnology alternatives to CMOS.

1.1 More Moore

This ultimate scaling of CMOS will be essential to supply the massive computing power and communication capability needed for the realisation of European Ambient Intelligence (AmI) applications at an affordable cost and a power efficiency ex-

[1] 2015–2018 is generically considered as the end of CMOS scaling because it has been shown that channel length will reach dimensions where the MOS device principle no longer operates.

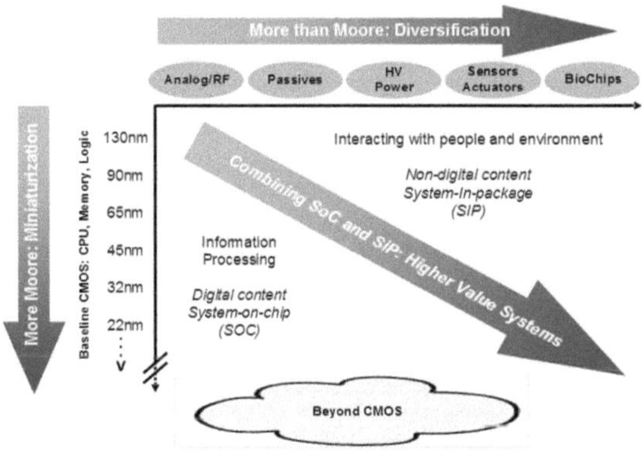

Fig. 2. Technology roadmap mirroring the European vision [16]

ceeding 200 GOPS/Watt for programmable and/or reconfigurable architectures [15]. However, reaching this ultimate CMOS node at the deca-nanometer level will require addressing cumulative interrelated challenges surveyed in [7, 8].

In the process technology domain, major challenges are:

- the massive introduction of new materials;
- the introduction of new device architectures;
- the move to EUV (Extreme Ultra Violet) lithography or nano-imprint lithography;
- the increase of random device and interconnect variabilities especially in memories;
- the limits of copper (Cu) interconnects (resistance, e-migration, cross-talk, etc.);
- the conflict between dynamic and static power.

In the design domain, the key challenge is the fact that ultimately non recurring engineering (NRE) cost may reach 1 B€/platform, if no drastic changes in design technology occur due to increased hardware-software, as well as technology-hardware interactions on multi-core platforms.

Perhaps even more relevant is the fact that process technology challenges directly impact design challenges. One relevant example for handheld terminals is the fact that static power will become significant in the energy consumption bill. Unfortunately, it is not expected that battery power density will evolve at the same pace (see Fig. 3). This has led to the introduction of several new design techniques, like power gating. On the technology side, the introduction of High K Metal gates has also greatly improved the gate leakage (not shown in the general trend curve below).

Random variability will affect parametric yield and will require novel ways to avoid corner-based design to cope with device uncertainty, and amenable to design automation. This will require the development of self-healing, defect- and error-tolerant, yet testable design styles based on low-cost on-chip adaptive control systems. The same adaptive control can be simultaneously used to track dynamically the best operating point, thus optimizing energy dissipation as well as thermal hot-spots.

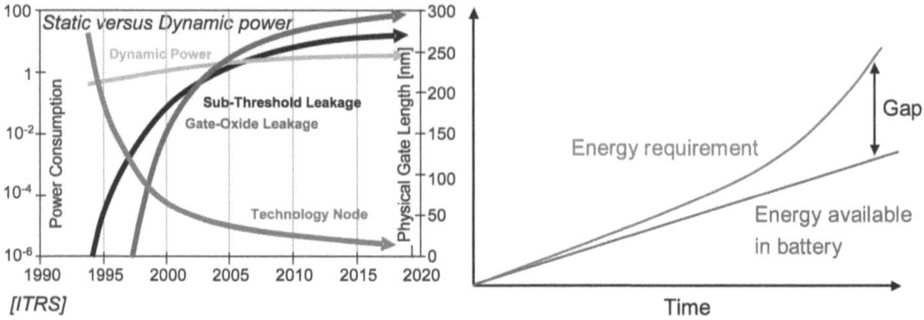

Fig. 3. Power consumption evolution with technology node vs battery evolution

Reliable local and global on-chip communication in 22 nm or smaller technology will be a much more limiting factor than transistor scaling and will require. Besides the investigation of optical, wireless or CNT-based technologies (i.e., in interlayer vias), investigation of architectural solutions such as tile-base Globally Asynchronous Locally Synchronous (GALS) architectures exploiting Networks-on-Chip and MP-SOC as global architectural mitigators. 3-D integration and SIP must also be studied as strong contenders to ultimate scaling for true system design, which is finally the ultimate goal of electronics.

Analog and RF design will have to cope with ultimate digital scaling and further sub-1 Volt scaling. This will require extreme creativity in analog and RF system design by compensating analog deficiencies through digital techniques. An alternative option will be to use a 3D assembly, exploiting an aggressive technology only for the digital part, while using a mature technology for the analog and RF functions.

Alternatives to bulk CMOS shall also be considered to overcome the shortcomings when scaling down to deep submicron. Partially or Fully Depleted Silicon On Insulator (PD, FD-SOI) technologies are foreseen as relevant with this regard, as it

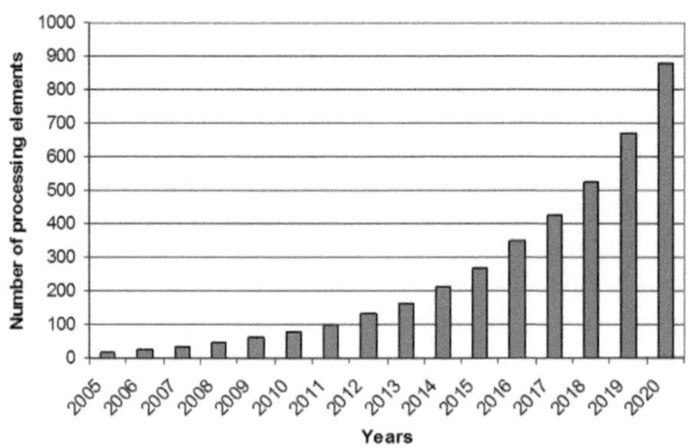

Fig. 4. Evolution of processor number in MP-SOCs

could lead to a better tradeoff between active and static power leakage. A significant active power reduction can be achieved by using SOI devices. Indeed, SOI devices are well known to be able to achieve the same performances than Bulk device, but with a lower power supply. This is achieved thanks to lower parasitic capacitances (thick buried oxide for electrical isolation instead of junctions) and thanks to lower threshold voltage in dynamic mode.[2] Active power reduction up to 50% can then be achieved with SOI. DC leakage can also be controlled [4]. For the very advanced technology nodes, FD-SOI technologies, thanks to their excellent channel electrostatic control offer a significantly improved ON/OFF current ratio. In addition, they offer lower intrinsic device variability.

1.2 More than Moore

The "More than Moore" approach intends to address parallel routes to classical CMOS by tackling applications for which CMOS is not optimal. These applications can be classified in three major categories: interfacing to the real world, enhancing electronics with non-pure electrical devices, embedding power sources into electronics [16].

In the field of IC design for advanced 3G standards, the More than Moore class is expected to bring significant breakthroughs in RF front end design. Indeed, new complex signal modulations (e.g. OFDM) require very linear RF components in order to limit distortion and ensure high signal throughput. Cellular phones can utilise up to seven different wireless standards or bands, including DCS, PCS, GSM, EDGE, CDMA, WCDMA, GPS and Wi-Fi, and each standard has its own unique characteristics and constraints. Additionally, next generation phones cannot be significantly larger than today's phones and they will need to have similar talk and standby battery lifetimes. Today, a large proportion of the components in a mobile phone are space-consuming "passive" elements such as inductors, variable capacitors and filter devices. Integrated passive elements and RF MEMS/NEMS have been proposed to help solve these problems. Key functionalities are filtering, time-base, tunability and reconfigurability. High-quality passive elements are available through System in Package (SiP) technologies. Besides, nano-materials are expected to strongly improve the achievable capacitance per unit area value for capacitors. Current research works are also exploring new materials towards bringing tunability and accordability to devices and circuits. One key remaining issue though, is whether the extra cost of these non-standard technologies can be justified by the benefits they bring. Thus, whenever heterogeneous technology is considered, the trade-off between performance and cost must be analysed.

In addition to those aspects, current mobile phones also embed many More than Moore technologies for their user interfaces: imagers, sensors, power devices...

[2] In case of partially depleted SOI devices.

1.3 Beyond CMOS

The "Beyond CMOS" paradigm intends to identify technologies that could replace CMOS, either in a disruptive or evolutionary way after CMOS will reach its ultimate limits. The ITRS Emerging Research Devices (ITRS-ERD) proposes criteria to evaluate the potential of emerging research devices and circuits with respect to future applications. The analysis presented in the ITRS-ERD document [12] is based on defining a set of criteria for logic and another set of criteria for memories, and applying them to potential technologies. These criteria are as follows:

- *for logic:* scalability, performance, energy dissipation, gain, operational reliability, operating temperature, CMOS technological and architectural compatibility;
- *for memories:* scalability, performance, energy dissipation, OFF/ON ratio, operational reliability, operating temperature, CMOS technological and architectural compatibility.

Nano-technologies falling into this category consider building blocks such as: Atom scale technologies, Spin electronics, Molecular electronics, Ferromagnetic devices, Nanoelectromechanical systems, Organic /plastic electronics, Bio-sensors. Because these new devices have behaviours that sometimes differ significantly to classical transistors, "electronics" using these functions needs to be invented as well. From an architectural viewpoint, these technologies derive into: bio-inspired electronics, nanomechanical computing, quantum computing. Stating this, it is obvious that a description of the research challenges related to Beyond CMOS is far too broad to be surveyed in this document. What can be kept in mind is that Beyond CMOS is extremely pluridisciplinary with extensions at all levels from building blocks to system usages.

One technology which is quite likely to develop and be used in Telecom devices is the Resistive Memory (RRAM). Phase Changed (PCRAM), as well as Magnetic (MRAM) memories are already in production; alternative technologies like CBRAM (Conductive bridging) or OXRAM (Oxide resistive) are under extensive research.

2 Processing needs in wireless communications

Bearing in mind the evolution of silicon technology, designers also have to consider the evolution of the requirements from advanced wireless systems. Along their path towards next generation broadband wireless access, different standardization bodies (e.g. 3GPP, 3GPP2, IEEE, ETSI-DVB, etc.) have been introducing new standards that enhance their legacy radio access technologies. Examples of recent releases are: 3GPP Release 7 (HSPA+), Release 8 (LTE), Release 9 (LTE-Advanced), 3GPP2 (UMB), IEEE WLAN 802.11 (n, vht), IEEE WMAN 802.16 (d, e, m), ETSI DVB (T2, H, SH, NGH). Emerging and future wireless communication systems are characterized by a clear and steady convergence on both services and technologies. From the service perspective, operators are striving to offer to the users a wide spectrum of rich multimedia services including both interactive and broadcasting, which raise the need to embed complementary technologies (e.g. for uni-cast, multi-cast, and broadcast transmissions) into the future generations of radio access systems.

This context of coexistence and convergence is driving demand for flexible and future-proof hardware architectures offering substantial cost and power savings. Manufacturers have already started activities towards the provision of multi-mode handsets featuring the advancements of the recent radio access technologies (e.g. 3GPP LTE, WiMAX IEEE 802.16m, DVB-T2/H). However a large gap is growing in the field of flexible radio between advances in communication algorithms, methods, system architectures on one side, and efficient implementation platforms. Despite the large amount of available results in the system level technologies related to multi-mode, multi-standard interoperability and smart use of available radio resources (adaptive coding and modulation, cross layer optimization,...), a very limited number of hardware solutions have been proposed to really support this flexibility and convergence by means of power efficient, low cost reconfigurable platforms. In most cases, chipset vendors offer different solutions for each combination of standards and applications to be supported.

To enable a single modem to service multiple different wireless systems, highly flexible solutions are needed. In practice, currently implemented flexible hardware modems are focused on the receiver segment placed between the RF front end and the channel decoder. In this part of the modem, several digital signal processing algorithms, such as equalization, interference cancellation multipath correlation (rake receiver), synchronization, quadrature amplitude mapping/demapping and FFT can be run on vector processors, which allow for giga-operations per second (GOPS) rates resorting to very high level of parallelism. However, other functional components of a modern modem (such as channel decoding) are not efficiently supported by vector processors and alternatives to software programmable architectures are not considered solid solutions: cost of hardwired dedicated building blocks becomes rapidly unacceptable with the number of standards to be supported, while reconfigurable hardware, such as FPGAs, are too expensive in terms of silicon area and standby energy consumption (which is due to leakage current, proportional to area).

The design of high throughput software programmable architectures is currently the largely prevailing development area for companies in the field of next generation base-band platforms. This dominating approach is the optimal solution for the current relatively limited needs of flexibility, but it is likely to be inadequate for future wireless systems that will be characterized by:

- significantly larger amounts of standards and communication modes to be supported;
- higher number of complex and heterogeneous processing algorithms;
- higher level of dynamic flexibility to support cognitive and opportunistic radio concepts;
- lower energy dissipation in both static and dynamic conditions.

Future of wireless communications cannot be guaranteed by current approaches towards flexible base-band platforms. Beside this research effort gap, known platforms for digital base-band processing show serious lacks of capabilities with respect to the radio flexibility that is currently studied at the system level and expected by the market. In particular presently available platforms suffer from two main limitations:

- *Partial flexibility*: since large difference is recognizable between the processing functionalities that are expected to be supported in a flexible receiver, and the actual level of flexibility that is achieved in hardware. As an example, advanced channel decoders are usually not included in the whole receiver as reconfigurable elements, but they are supported by means of separate components, designed and optimized to execute one specific decoding algorithm.
- *Expensive flexibility*: since flexibility comes at a very high cost in terms of occupied area and dissipated energy and required reconfiguration time; on the contrary, flexible platforms are requested to provide better overall die area and power figures than receivers designed by simply allocating multiple function-specific components. Moreover simple, fast and energy efficient reconfiguration procedures are of primal importance to enable true flexibility, as awaited in cognitive and opportunistic radio systems.

2.1 The need for highly demanding building blocks

A rather reduced number of computationally intensive functionalities is associated with the key enablers commonly adopted by next generation radio access standards. A list of these technologies is presented in Fig. 5. The table on the left shows key characteristics of FEC technologies adopted in several standards that have been introduced during the last 15 years in the domain of wireless communications and digital broadcasting. Particularly the table contains for each specified FEC the maximum data throughput and the processing complexity, expressed in GOPS. The plot on the right side of Fig. 5 clearly indicates that both throughput and complexity tend to increase exponentially with time. More in details, data throughput doubles within 15 months and this trend is almost in agreement with the evolution trend of performance in semiconductor industry, at least until Moore's law remains valid (see Section 1). However, Fig. 5 shows that the FEC complexity trend is faster, as the required GOPS doubles every 12 months.

The growing gap between silicon performance and FEC complexity trend lines implies that the efficient implementation of emerging and future error correcting techniques will not be guaranteed by the progress in the semiconductor industry,

Standard	FEC	rates			
GSM	CC	3/4 - 1/4	33 - 876	up to 12 kbps	0,001
EDGE	CC	6/7, 1/3	39 - 870	5 - 62 kbps	0,005
UMTS	bTC	01-mar	40 - 5114	up to 2 Mbps	3
HSDPA	bTC	1/2 - 3/4	40 - 5114	up to 14.4 Mbps	30
CDMA-2k	bTC	1/2 - 1/5	378 - 20736	up to 2 Mbps	0,2
(WiMax)	dbTC	1/2 - 3/4	up to 648	up to 54 Mbps	72
IEEE802.16 (WiMAX)	LDPC	1/2 - 3/4	up to ~2500	>100Mbps	7
DVB-T/H	CC	1/4 - 7/8	(broadcast)	up to 32 Mbps	10

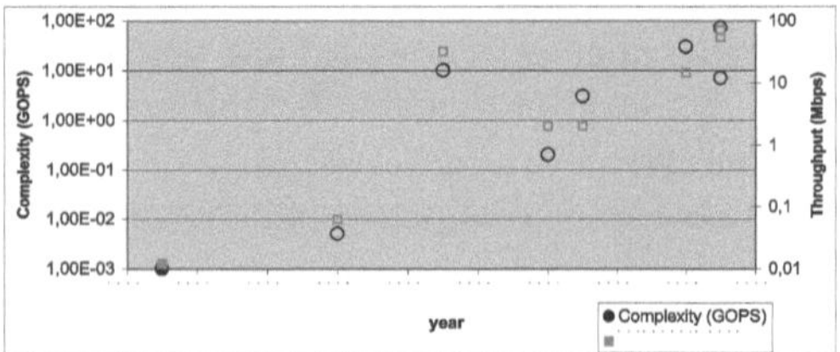

Fig. 5. Throughput and complexity trends for wireless standards

but it will continue to impose application specific optimizations at the confluence of algorithm and architecture. Particularly in the implementation of computationally intensive base-band processing tasks, this joint effort at the algorithm and architecture levels can be targeted towards different optimization objectives, such as area occupation, throughput, power dissipation, and flexibility.

As for the area and throughput objectives, several cases can be cited to show how efficient implementations often come from joint design efforts spent at the algorithm and architecture levels. For example, although the original LDPC decoding algorithm requires rather complex processing at the check nodes, all decoders implemented in the last few years resort to the min-sum approximation, which saves a significant portion of complexity with a marginal performance loss. Another relevant example is found in the MIMO detection domain, where several sub-optimal implementations have been proposed in the last few years as alternatives to sphere decoding: these solutions (e.g. k-best and LORD) exhibit close to Maximum Likelihood performance, simpler architecture and deterministic detection delay at the same time. Additionally the algorithm-architecture interdependency in a specific application domain can be leveraged to improve energy efficiency.

Joint algorithm and architecture optimization can also be seen as a method to achieve flexibility through the development of unified processing algorithms that enable increased sharing of hardware resources. This unifying approach combines both algorithm and architecture alternatives into a joined optimization effort towards efficient flexibility in radio communications. The idea is to develop innovative processing algorithms, where limited and controlled performance degradation is accepted and exchanged for flexibility in the hardware implementation.

2.2 The need for power efficient design

Several IC implementations of turbo and LDPC decoders have been published in IEEE ISSCC (International Solid State Circuits Conference) and JSSC (Journal on Solid State Circuits) series in the past few years. Looking at the characteristics that have been measured for those components, one can easily see that the power dissipation trend-line remains fairly constant across the last ten years and typically included in the range between a few tens and a few hundreds of mW. This general tendency is somehow surprising, as throughput and processing complexity have been increasing along the same period of time. The explanation of the observed trend on power dissipation comes from CMOS process down-scaling (particularly scaling of gate area and parasitic capacitance) that has substantially balanced the increasing in computational effort. This appears as a very bright and encouraging conclusion. However power dissipation is expected to become a very critical issue in future developments for several reasons.

First of all, the static power dissipation has been usually neglected so far; however it will soon become comparable to the dynamic one in next generations of CMOS process technology (see Section 1). Therefore, with a fixed power budget assigned to base-band components, the dynamic power consumption will need to be limited to a lower bound than in today's implementations.

Secondly, some base-band functions will increase dramatically the need of processing energy. The most relevant example is probably given by joint MIMO detection and channel decoding. The concatenation of soft output MIMO detection and iterative error correction algorithms will create high complexity receivers where the FEC processing is organized around two nested feedback loops: an inner loop, associated with turbo or LDPC decoding iterations, and an outer loop, with exchanged soft information between MIMO detector and inner channel decoder. This arrangement will significantly increase the global complexity and affect both the required processing speed and dissipated power. For example, if a 300-mW turbo decoder is used as the inner unit in a concatenated system with 3 iterations of the outer loop, the power consumption of the turbo decoder will increase by a factor of 3.

A third reason comes from the increasing levels of flexibility that will be incorporated in next generations of channel decoders. Flexibility is ineluctably associated to a cost in terms of consumed energy, which tends to reduce the power efficiency, although recent work has shown that in-depth architecture/algorithm analysis and design can significantly reduce this overhead [2, 20].

The rise of power consumption combined with the wished reduction in size of handset devices causes temperatures to increase because the transfer of heat is proportional to the surface area. Increased temperatures have two effects. The first is that the temperature of the casing of the device can go so high that it becomes too hot to handle. The second effect is that higher temperatures make the electronic components unreliable and more likely to fail. In [9] it is envisaged that a dramatic increase of energy consumption of 4G mobile device will make active cooling a necessity, which is not attractive for users and manufactures. In [27] the performance of active cooling in mobile devices is investigated. From the mobile manufacturer's perspective the energy consumption problem is critical, not only technically but also taking into account the market expectations from a newly introduced technology. This is in fact becoming a key concern: there exists a continuously growing gap between the energy of emerging radio systems and what can be achieved by:

- Battery technology evolution;
- Scaling and circuit design progress;
- System level architecture progress;
- Thermal and cooling techniques.

Considering this power consumption (and dissipation issue), designers are more and more considering power consumption as a key figure of merit at design time. In Digital Signal Processor (DSP), the MIPS/mW is a metric that has been used for a long time. Similarly, communication building blocks efficiency is evaluated in terms of performance vs. energy consumption trade-off. In channel decoders for instance, power efficiency η is defined as the ratio between the number of decoded bits per second and the corresponding dissipated power. It is usually measured in Mbps/mW. The numerator of the ratio can be written as $N_b \cdot f_{\mathrm{ck}}$, where N_b is the number of bits decoded per clock cycle and f_{ck} is the clock frequency; neglecting the static contribution, the denominator can be expressed as $A \cdot C_{\mathrm{sw}} \cdot V_{dd}^2 \cdot f_{\mathrm{ck}}$, where A is the total occupied area, C_{sw} is the average switched capacitance per unit area, and V_{dd} is the

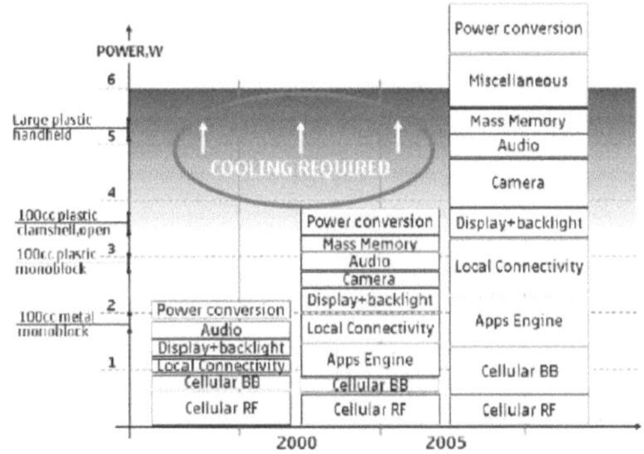

Fig. 6. Power consumption increase in wireless terminals [9]

supply voltage. Defining $A_b = A/N_b$ as the average area required to process one bit, the power efficiency can then be formulated as $\eta = 1/(A_b \cdot C_{sw} \cdot V_{dd}^2)$. This simple model clearly shows that the power efficiency only depends on three parameters: two of them, C_{sw} and V_{dd}, are tight to the evolution of silicon technology (see Section 1), while A_b largely depends on how the required processing functions are implemented. When two or more functions are mapped to a unique architecture capable of flexibly supporting all of them, this leads to the allocation of additional components that are used to handle the switching overhead between two functions. This results into an increased average area per decoded bit and impairs the efficiency. Therefore the search for efficient flexibility is the search for architecture solutions that minimize A_b.

2.3 *The need for flexibility*

Although throughput and area have been the dominant metrics driving the optimization of digital building blocks, recently, the need for flexible systems able to support different operative modes, or even different standards, has changed the perspective. In particular, the Software Defined Radio (SDR) paradigm made flexibility a fundamental property of future receivers, which will be requested to support a wide range of heterogeneous standards.

This is a very ambitious and innovative challenge that shall provide support to multiple versions of a specific functionality, each one characterized by a different trade-off between communication performance and energy (or throughput) efficiency. The fundamental purpose here is to dynamically enable the change between one version and another of the considered functionality, in response to energy constraints and user needs. This type of versatile platform will therefore give support to complex power management algorithms and optimal allocation of the spectrum resources. Although the concepts of adaptive and cross-layer optimization in mobile terminals are not new, the design of an implementation platform supporting those concepts is still an open problem and a very challenging research topic. Two key problems can be seen: the required level of flexibility is higher than in classical multi-standard architectures, and constraints on the reconfiguration latency are expected to be stricter.

Flexible algorithms and architectures must be developed to enable the support of energy management techniques involving all functionalities of the digital base-band processing chain. Specific optimization metrics and methods need to be introduced to drive algorithm and architecture design. Proper methods must also be developed for estimating the operative conditions and algorithms for realizing the energy management.

A relevant application example for the mentioned flexibility target is given by next generation wireless systems that use multiple antennas to deliver very high data rate services. In such systems, a feedback loop between MIMO detector and outer channel decoder enables iterative "Turbo-MIMO" processing: performance very close to the A Posteriori Probability (APP) detection has been achieved with different detection techniques. While optimum error rate performance is obtained with soft-output maximum likelihood detection, linear and successive interference cancellation (SIC) algorithms are interesting alternative solutions, with an implementation complexity lower than the sphere decoding. Moreover, for low code rates or low modulation order, plain linear detection performs within 2dB of the performance bound; on the other hand advanced detection strategies (sphere decoding) is convenient for high code rate and higher order modulation. Thus different performance-complexity-energy trade-offs are covered by a set of heterogeneous algorithms and a proper flexible platform is required to dynamically exploit the energy minimization opportunities offered by these trade-offs.

Current approaches to multi-standard functionality pragmatically aim at implementing a "just enough flexibility", by supporting codes and throughput requirements specified in some of the current standards. A rather small number of multi-standard decoders have been implemented so far. They can be classified in three categories.

- *Intra-family flexibility*: which support multiple modes belonging to the same functionality. As an example, one could design a turbo code decoder able to operate over several turbo codes, specified in different standards, such as UMTS, WiMAX and WiFi. The most common implementation approach for this category is hardware parameterization: the processing architecture is organized around a number of storage and computation units that are structured based on a number of parameters, such as block size and code rate.
- *Inter-family flexibility*: which is capable of a wider flexibility, as they must process functions belonging to different and in some cases heterogeneous families (e.g. to stay with the FEC example, turbo and LDPC codes could be an example) specified in two or multiple standards. In this case, reusable hardware resources can be identified and shared among supported codes, provided that hardware sharing results into saved area with respect to the straightforward allocation of several independently designed decoders ("Velcro approach").
- *Full flexibility*: which supports high throughput implementation of a wide range of heterogeneous function, not necessarily limited to the ones that are today specified in a standard. As an example, a fully flexible turbo code decoder should be

able to support any interleaving law. The additional difficulty of this approach derives from the fact that parallel collision-free decoding architectures are heavily based on the specific features of the code family to be decoded; as a consequence, these architectures can hardly be exploited when multiple different codes must be supported. Common operator technique is another approach belonging to this class [2].

3 Software defined radio approach: Challenges and opportunities

Flexibility requirements, when coupled with low-cost and less time-to-market constraints, make the development of a mobile device highly complicated and challenging. SDRs, with cognitive capabilities, are getting prominence as potential candidates to meet the future requirements of mobile devices. Compared to the pragmatic design approach for flexibility motioned above, the SDR approach aims at providing a comprehensive design framework encompassing platforms, architectures, software, methodology and design tools.

3.1 Current solutions

Current solutions for SDRs are component based and model driven where a Platform Independent Model (PIM) of a waveform[3] is constructed as an assembly of components, e.g. [24], from a specification document, e.g. [17]. Each component represents a part of functionality in a whole waveform. From a PIM model, a Platform Specific Model (PSM), which denotes the implementation of a waveform, is obtained with/without using a library. Libraries internally developed or from third party vendors providing efficient implementations of few, sometimes even all, components of a waveform can improve overall system efficiency and drastically decrease development time. For example, Texas Instruments (TI) provides efficient implementations for implementing the WiMax waveform [28].

Some other approaches for developing SDRs are based on Software Communications Architecture (SCA) proposed by JTRS [13]. SCA is predominantly General Purpose Processor (GPP) based and uses CORBA as middleware abstracting the underlying hardware. This creates an operating environment that enables to develop applications independent of hardware, and methods for loading new applications, configuration and control.

3.2 Key issues

Though each of the above solutions improves development of SDR in one way or the other, there are associated issues often leading to situations where one solution does not fulfil all requirements. A serious drawback in the libraries that are available

[3] In this context, the term waveform represents a complete wireless standard with several modes.

today on the market is that they are specific to one waveform or to one hardware platform. For example, the library in [28] is specific to the WiMax waveform targeting a specific Processing Element (PE), namely the TMS320TCI6482 DSP. Moreover, some of the libraries are proprietary in nature; the details on the components and their interfaces are not known. Even though library based approaches have the potential to increase efficiency and portability, the lack of standardization decreases reuse of implementations drastically.

Similarly, overhead caused for supporting Service Component Architecture (SCA) is a key deterrent for its usage, particularly in physical layer processing, due to the existence of hard constraints, e.g. latency [26]. Though abstraction of the underlying hardware platform makes mapping of a waveform description onto a hardware platform easy, it completely blocks the opportunity to exploit the architectural capabilities of a hardware platform. Hence, an optimum mapping is not possible. Mapping should consider the requirements of a waveform, e.g. processing complexity, available resources in a hardware platform, e.g. memory, and constraints of a waveform, e.g. throughput, when optimizing with respect to design requirements like energy efficiency. Therefore, constraint aware mapping is a key for improving the overall efficiency of the complete system.

The efficiency of a waveform implementation is a pivotal factor for overall footprint and energy efficiency. Investigations done on implementation efficiency indicate optimization limits depending on the implementation type [29]. For example, implementing in assembly code is more efficient than C-code because assembly code can better exploit the architecture of a PE. A GPP offers high flexibility, but requires more energy per decoded bit than, e.g. a Digital Signal Processor (DSP). Therefore, entirely GPP and C-based SDR solutions are not suitable for battery driven devices due to low implementation efficiency.

Merely increasing parallelism in order to increase computation power, without considering efficiency will lead to high area and energy consumption. Therefore, in SDR systems where future requirements of computational performance will be in the order of tens or thousands of Giga Operations Per Second (GOPS), techniques like massive pipelining, increasing the number of General Purpose Processor (GPP) cores or speeding up the clock are not satisfactory due to energy efficiency reasons. HW platforms for SDRs will, most likely, be heterogeneous in nature with programmable PEs like Application Specific Instruction-set Processors (ASIPs), DSPs, GPPs and Application Specific Integrated Circuits (ASICs) or even physically optimized ICs. Design requirements of a SDR system, including flexibility and efficiency, will determine the type and number of PEs. For example, physically optimized ICs provide very high performance and power/energy efficiency; however they offer least flexibility whereas GPPs provides full flexibility at the cost of very low energy efficiency.

Numerous issues in waveform development for SDRs are due to the "specification to implementation" problem. In general, waveform specification is in the form of a textual document with details on different modes, constraints and critical loops that have to be met by a waveform implementation. Textual documents provide redundant information, which is sometimes verbose and sometimes terse. Therefore, creating a PIM of a waveform incorporating all the features like latency and deriving a PSM model, meeting key requirements like throughput from it, is a cumbersome task, often error prone. Therefore, design, development, integration and testing of waveforms have become highly complex and time consuming.

3.3 Challenges

The paradigm of SDR poses new challenges or makes current design challenges more stringent. The most relevant ones are:

- *Portability*: which can be defined as the inverse of porting effort, represents the ease with which one waveform can be moved to another hardware platform [29]. Portability requires a platform independent waveform description.
- *Efficiency* with respect to area and energy is essential in order to decrease the power/energy consumption and extend the battery life. However, this requires high efficiency in waveform implementation.
- *Interoperability* denotes the ability that a waveform implemented on two different hardware platforms interoperates with each other.
- *Loadability* illustrates the ease with which a waveform can be loaded, over-the-air, into a hardware platform, programmed, configured and run. Loadability can be increased by well defined and known interfaces in waveform implementation.
- *Trade-offs* between flexibility and efficiency becomes challenging in the wake of their contradictory nature. This makes heterogeneous Multi-Processor System-on-Chips (MPSoCs), an inevitable candidate as the hardware platform for implementing a waveform.
- *Cross layer design* and optimization techniques are getting popular, if not mandatory, in order to cope with the increasing need for spectrum and energy efficiency. This leads to very tight dependencies, interactions between physical and MAC, higher layers that have cognition, requiring flexibility in implementation and algorithms.

Most of the challenges in SDR arise due to the contradictory requirements of flexibility, performance and efficiency. Heterogeneous MPSoCs with specialized PEs can pave the way to solve the dilemma of contradicting demands of high computational performance at the one hand and energy efficiency on the other. However, designing such a system is a challenging task. Tools are required for the development of the dedicated PEs as well as of the whole SoC. High speed simulation is necessary in order to support design space exploration and verification at an early phase.

Still, the complexity of modern, flexible implementation structures would hardly be manageable and their development is a tedious and error-prone task. What is needed is a description method that can lead to a (semi-)automatic generation of a waveform implementation directly from the specification. Therefore, a methodology is required, that raises the abstraction level of receiver design to make it manageable.

3.4 Opportunities

As mentioned earlier, direct implementation on a low abstraction level is not well suited for an efficient portable waveform implementation. Raising the abstraction level leads to library based approaches, where efficient implementations of basic components are available and can be assembled to implement the complete transceiver. Also, a library based approach enables efficient utilization of heterogeneous MPSoCs.

3.4.1 Design principles

There are several key design principles that must be considered while building a library that can pave way for Waveform Description Language (WDL) based SDR development. They are:

- Hardware architectures that offer full flexibility, e.g. GPPs, are not efficient and are costly in terms of area and energy consumption. Therefore, application specific optimization is needed in order to increase both energy efficiency and computation performance. If limited flexibility can be offered, such architectures can still be tuned for different requirements.
- Algorithms that might work efficiently for one scenario might not be efficient for another, e.g. sophisticated and complex algorithms might be needed in a bad channel while simple algorithms might be sufficient in a good channel. This creates the need for analyzing algorithms that are scenario-specific and to identify the common kernels in these algorithms ("Nuclei") to maximize reuse.
- Building a library that is based only on functionalities in waveforms limits reuse of the library. For example, if one of the components in a library is a modulator of a particular scheme, a different scheme in another waveform renders it useless. Therefore, emphasis should not be on the functionalities but on the algorithms that are used for implementing such functionalities. This not only increases reusability, but also provides algorithmic flexibility.
- Flexibility in implementing a waveform can be provided, even in a fixed hardware platform, through different implementation algorithms and configuration parameters like implementation-method, input data-width, scaling, etc. However, a PE in a hardware platform should have architectural capabilities to support different implementation algorithms efficiently.
- Providing easy programmability for complex systems like SDR is essential to exploit efficiently the hardware resources. A programming model that can bridge the gaps between waveform, hardware platform and mapping is needed. This model should allow a designer to utilize the flexibility present in a hardware platform in order to increase the implementation flexibility.
- Due to the presence of a number of layers with very high interaction between them in typical waveforms, it is essential to treat SDR development as a joint optimization problem.
- In spite of advances in standardization due to bodies like SDR Forum [25], NGMN alliance [21], JTRS program [13], etc., lack of complete and unified standardization is preventing huge advances in SDR technology. As a consequence,

reusability of other solutions, participation of different vendors is limited, indirectly leading to increase in development costs. This also prevents co-operations and sharing knowledge between academia and industry on the one hand and between military and civil domains on the other hand.

Due to the strong dependencies between algorithms, hardware architecture and tools, it is necessary to investigate these aspects jointly in order to identify an efficient SDR development methodology. For example, a detailed algorithm analysis can drive the component identification and implementation, which then feeds back the analysis results which may cause revision of the algorithm itself, making it algorithm architecture co-design. Furthermore, the real implementation of Nuclei has the potential to deliver important information on the interfaces and parameters which are required for tools exploiting the spatial and temporal mapping of a waveform description. Therefore, joint results achieved by working together in the three domains listed before are needed, making SDR development algorithm, architecture and tools co-design.

3.4.2 Algorithms

Since SDRs have to offer flexibility, it is efficient, if not necessary, to exploit the tradeoffs between complexity and error rate performance in different algorithms. For example, in a spatially multiplexed MIMO signal, though exhaustive search delivers the minimum error rate, the enormous computational complexity is a heavy burden for the base band receiver. On the other hand, low-complexity algorithms such as zero-forcing detection can operate in a limited SNR range only. Therefore, it is essential to analyze such algorithms jointly along with their trade-offs.

Algorithms that are used for implementing different functionalities can have common computation and communication patterns. This commonality can be exploited by identifying such common kernels that are also computation intensive. Such algorithms might be used in different applications. If the granularity of such kernels is optimum, i.e. not coarse grained as a complete channel decoder nor as fine grained as an adder, it can enhance reusability and can enable the availability of optimized implementations for such kernels. However, emphasis should be on the implementation-friendly algorithms in order to enable various implementation alternatives, based on different algorithms, without sacrificing efficiency.

3.4.3 Tools

Tools must offer a seamless environment for developing SDRs and then Cognitive Radios by providing the infrastructure to capture the waveform specification, to do mapping, implementation, integration and verification. Due to a huge number of critical paths involving several components of a waveform, a constraint aware mapping approach is needed. It increases the chances of successful mapping and decreases the number of iterations. However, it complicates not only the tool development but also identification of the appropriate ways for describing the impacts with respect to constraints. Hence, the complexity of wireless equipment of the future is particularly increased by the flexibility features. On the one hand this induces a permanent mix of hardware and software design all along the design cycle, which requires the use of mixed representations, enabling to model the design at several levels of granularity. On the other hand flexibility also implies a large number of possible execution scenarios. As a consequence, new software environments and design approaches are needed to represent at each design steps the suitable level of abstraction. The solution to such challenges is to use a high level design approach [19] throughout the design steps. This would permit to simplify the un-necessary details at each level of interest, leading at the end of the conception to the usual current design tools for the last implementation steps. In particular, at the very early design stage, this may help dealing with the variety of use-cases expected for the equipment without taking into account and waiting for the development of the entire system, such as proposed in [10] for cognitive radio design. Such a design paradigm means using modeling and meta-modeling approaches [14] associated to automatic transformation [5] in order to go through the design flow step by step.

One of these steps in particular aims at validating and evaluating the spatial and temporal mapping decisions as well as the performance of the overall system. At this level, the environment enables to simulate the system behaviour in terms of functionality and timing. The software based system simulation plays a key role in exploration and verification of the spatial and temporal mappings. The information obtained by the simulation can be fed back to the higher layers in order to improve the mapping quality. This can be considered as iterative approach, which is repeated until a satisfying result is obtained. This example should be followed at all other necessary levels of the design.

3.4.4 Architectures

In general, heterogeneous MPSoCs can provide high performance due to parallel processing of tasks and at the same time provide flexibility and efficiency due to the heterogeneous, function-optimized nature of PEs. Among the PEs, ASIPs are very attractive candidates for implementing SDR systems, where a fine balance between flexibility through programmability and efficiency through application specific architecture optimization is essential. For example, conventional load store memory architectures may not be able to meet the throughput-latency demands of SDR applications and may become a bottle neck. Therefore, special application specific memory architectures, in addition to other architectural options, are needed to meet these demands. Similarly, due to extremely high throughput and short latency demands in the communication between PEs in an MPSoC, conventional communication schemes like buses are most likely to fail. Instead, the idea of specialized communication architectures, including dedicated links between PEs, even special links that are optimized separately for throughput and latency is gaining more interest.

To summarize, requirements and therefore complexity of SDRs are increasing day-by-day, mainly driven by new applications and services in wireless communication systems. Design and development of a SDR has inherently numerous challenges due to the contradicting nature of flexibility and efficiency requirements. However, this provides tremendous opportunities and calls for a radical change in the way such complicated systems are built. One promising approach, like [23], that has the potential to provide implementation flexibility even in a fixed hardware platform, is the Waveform Description Language (WDL) based development using a library of algorithmic kernels. This approach promises not only the participation of vendors by standardization and open interfaces, but also provides algorithmic and implementation flexibility even in a fixed hardware platform.

4 RF trends in flexible radio

The digital communication research on multi-standard radio has started based on the assumption of the Software Radio, which extrapolated that RF stages of a radio would be transparent for the baseband processing either thanks to highly flexible RF components or to very high speed converters. Both have shown limitations and further research is needed to achieve highly flexible Software Defined Radio. In fact, too major approaches emerge for designing a flexible RF. The first one considers very large band RF that can therefore accommodate several systems and bands. This approach suffers from bad sensitivity level though. The second one relies on tuneable components with which parameters can be adapted to match the system requirements. These rules often contradict the guidelines RF designer are used to consider when defining an RF architecture which is usually optimised for sensitivity, power consumption, and IC integration.

At the transmitter side, classical approaches usually result in low flexibility architectures sketched in Fig. 7. In such designs, lump elements freeze the circuit performances to given specifications. Thus, this classical circuitry will hardly be adaptable. Multi-standard terminals based on this concept end up with a RF front-end comprising several RF ICs in parallel, also referred to as the Velcro approach. In order to come up with a less costly and bulky approach, new architectures in which the boundary between the analog and digital worlds has been modified to enable the use of waveform shaping in the digital domain has been proposed.

For instance [30] suggests using a LInear amplifier with Non linear Components (LINC) architecture to efficiently address large Peak to Average Power Ratio (PAPR) OFDM signals. The advantage of this approach is that amplifiers have to handle constant envelop signals despite the non constant envelop of the OFDM signal. This leads to better power efficiency and better flexibility, especially when the signals are shaped in the digital domain [22]. Despite its higher flexibility, this

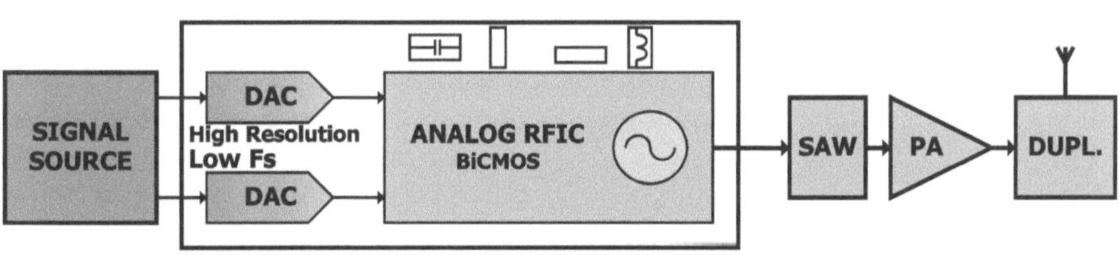

Fig. 7. Classical TX architecture

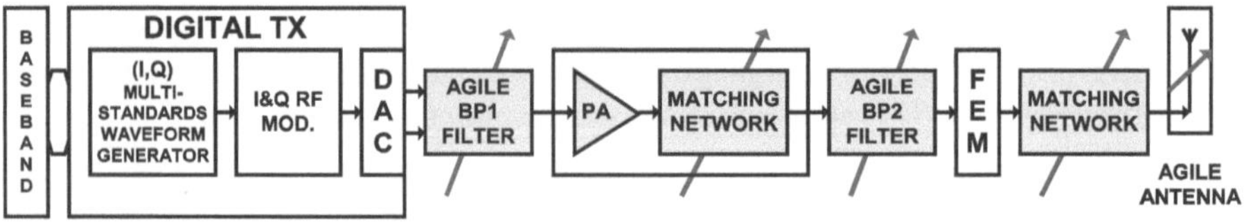

Fig. 8. Flexible TX architecture

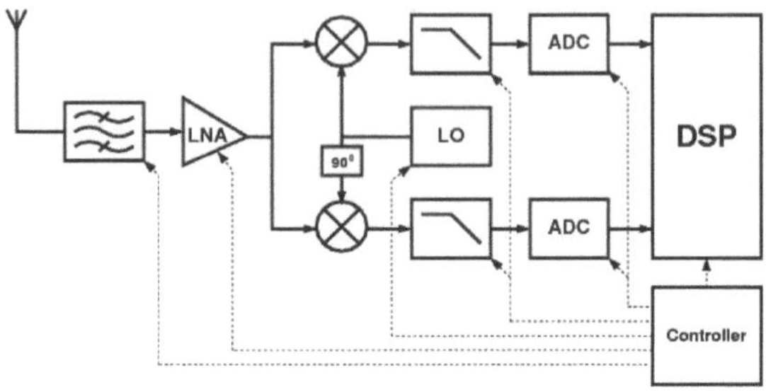

Fig. 9. Flexible zero-IF RX architecture

architecture still hardly copes with wide band signals and cannot be tuned over a large central frequency range. This is mainly due to the limited flexibility offered by nowadays analog stages.

Similarly, the trend at the receiver side is to limit the number of analog components. The hype for zero-IF or low IF architectures in the past few years partly came from this trend.

The zero-IF architecture (Fig. 9) has indeed several fundamental advantages over its heterodyne counterpart. The intermediary IF stages are removed and the functions of channel selection and subsequent amplification at a nonzero IF are replaced by low-pass filtering and baseband amplification, for which a monolithic integration is feasible. Although zero-IF exhibit relevant specifications, it suffers from well identified problems such as DC offset, LO leakage and I/Q mismatch, the first being the prominent one. The low-IF receiver concept has been developed to avoid these drawbacks. Fundamentally the low-IF receiver originates from the conventional heterodyne receiver system. The main difference is that the digitization process is shifted from the baseband part to the IF part. By implementing A/D conversion at this earlier stage, more flexibility at the receiver can be achieved. The concept of low-IF has become even more attractive recently, especially for emerging systems which require higher transceiver flexibility while keeping the terminals compact size and energy efficient. There are several benefits that can be obtained by implementing early conversion, namely: the high degree of programmability at the receiver, and the avoidance of issues associated with analog baseband demodulation, such as I/Q imbalance, DC offset, etc. Despite all these supporting facts, there is still one main obstacle for implementing such architectures, for it requires a fast high-bandwidth high-dynamic-range conventional ADC for converting radio signal with sufficient fidelity. Therefore, improving the performance of ADC is crucial to enhance the flexibility of RF receivers. Besides the performance, power consumption of the conversion stages is a matter of concern to integrate such solutions in low power battery operated devices.

Even when limiting the analog part of the transceiver, the use of tuneable filtering is needed, each filter being dedicated to the given bandwidth of the targeted system. RF filtering has always been considered as the bottleneck of the front-end implementation and making it tuneable represents a huge challenge [18]. For instance, Bulk Acoustic Wave (BAW) filters are highly selective band-pass filters that are convenient for a particular application. However, even if BAW filters are tuneable in frequency, this is only limited to a few percents and tuning control is quite complex to implement in practice. Besides, practical implementations based on Yttrium Iron Garnet (YIG) resonators provide multi-octave bandwidths and high quality-factor resonators. However, they consume a significant amount of dc power (1 to 3 W), and their linearity is poor. Moreover they are bulky, expensive and cannot be easily miniaturized for wireless communications. Alternatively, diode varactor-tuned circuits are simple and require little bias current and size, but they have not met the expectations in terms of loss. Solid-state varactors can provide a wide tuning range, but they have loss and linearity problems at microwave frequencies. Therefore, low cost and high performance tuneable solid state resonators is a myth. Besides solid state solutions, RF MEMS can provide a relevant alternative. Being constructed entirely of low loss metals and dielectrics, these mechanical structures feature inherently low loss properties.

5 Conclusions

Recent trends in silicon technology and communication system demands exhibit a growing gap between application needs and what the technology can deliver. A key driver for the telecom industry is the mobile business. Mobility, which relies on battery operated handheld devices, provide stringent requirements on equipments in terms of processing power, power consumption and flexibility. At the same time the battery and silicon technology does not progress at the same pace. The emergence of new standards implementing ever more efficient air interfaces also put stringent constraints on the design time. Thus, the reuse of hardware building blocks and a proper methodology and tools are needed to evaluate hardware performance tradeoffs at the earliest stage. Flexible radio is a promising approach in this regard. However, a unified framework is still to be found to enable the co-design of communication functions and hardware platforms.

References

1. Moore G.: Cramming more components onto integrated circuits. In: Proc. IEEE **86**(1), pp. 82–85 (1998).
2. Alaus L., Noguet D., Palicot J.: A reconfigurable LFSR for tri-standard SDR transceiver, architecture and complexity analysis. In: EUROMICRO Digital System Design (DSD), Parma, Italy (2008).

3. Alaus L., Palicot J., Roland C., Louet Y., Noguet D.: Promising technique of parametrisation for reconfigurable radio, the common operators technique: Fundamentals and examples. Journal of Signal Processing Systems (2008).
4. Belleville M., Faynot O.: Evolution of deep submicron bulk and SOI technologies in low power electronics design. In: Piguet C. (Ed.): Low-Power Electronics Design. CRC Press, Boca Raton, FL (2004).
5. Bondé L., Dumoulin C., Dekeyser J.L.: Metamodels and MDA transformations for embedded systems. In: Proc. Languages for Formal Specification and Verification, Forum on Specification & Design Languages, Lille, France (2004).
6. Cavin R.K., Zhirnov V.V.: Future devices for information processing. In: Proc. Eur. Solid-State Device Research Conf., Grenoble, France (2005).
7. Chen T.C.: Where CMOS is going: Trendy hype vs. real technology. ISSCC 2006 Digest of Technical Papers, pp. 22–28 (2006).
8. Declerck G.: A look into the future of nanoelectronics. In: Proc. Symp. VLSI Technology Digest of Technical papers, Kyoto, Japan (2005).
9. Fitzek F.H.P., Katz M., Zhang Q.: Cellular controlled short-range communication for cooperative P2P networking. Wireless Personal Communications (2008).
10. Godard L., Moy C., Palicot J.: An executable meta-model of a hierarchical and distributed architecture management for the design of cognitive radio equipments. Annals of Telecommunications **64**(7-8), pp. 463–482 (2009).
11. Hutchby J.A., Bourianoff G.I., Zhirnov V.V., Brewer J.E.: Extending the road beyond CMOS. IEEE Circuits and Devices Magazine **18**(2), pp. 28–41 (2002).
12. International Technology Roadmap for Semiconductors. [Online]. Available: http://www.itrs.net.
13. JTRS. [Online]. Available: http://sca.jpeojtrs.mil/.
14. Lecomte S., Guillouard S., Moy C., Leray P., Soulard P.: A co-design methodology based on model driven architecture for real time embedded systems. Mathematical and Computer Modelling Journal, no. 3-4, pp. 471–484 (2011).
15. De Man H.: Ambient intelligence: Gigascale dreams and nanoscale realities. ISSCC 2005 Digest of Technical Papers, pp. 29–35 (2005).
16. Report from the CATRENE scientific committee: Towards a "More-than-Moore" roadmap
17. Department Of Defense Interface Standard, Interoperability and performance standards for data modems. Tech. Rep. MIL-STD-188-110B, Department Of Defense, USA (2000).
18. Morche D., Belleville M., Delaveaud C., Ktenas D., Mayrargue S., Po F.C.W.: Future needs in RF reconfiguration from a system point of view. In: Proc. IEEE Bipolar/BiCMOS Circuits and Technology Meeting, Capri, Italy (2009).
19. Moy C.: High-level design approach for the specification of cognitive radio equipments management APIs. Journal of Network and Systems Management **18**(1), pp. 64–96 (2010).
20. Naoues M., Noguet D., Louët Y., Grati K., Ghazel A.: An efficient flexible common operator for FFT and Viterbi algorithms. In: Proc. IEEE Veh. Technol. Conf., Budapest, Hungary (2011).
21. Next Generation Mobile Networks (NGMN) Alliance. [Online]. Available: http://www.ngmn.org/.
22. IST-ORACLE, Overall complexity guidelines of an OR Terminal. Tech. Rep. (2009).
23. Ramakrishnan V., et al.: Efficient and portable SDR waveform development: The nucleus concept. In: Proc. Military Commun. Conf., Boston, MA (2009).
24. Rodriguez A.: Defense Communications: US MIL-STD-188-110B Waveform Simulation, July 2003. [Online]. Available: http://www.mathworks.com.
25. Software Defined Radio. [Online]. Available: http://www.sdrforum.org/.
26. Singh S., et al.: SCA based implementation of STANAG 4285 in a joint effort under the NATO RTO/IST panel. In: SDR Technical Conf., Washington, DC (2008).
27. Walsh E., Grimes R., Walsh P.: The performance of active cooling in a mobile phone. In: Int. Conf. Thermal and Thermo mechanical Phenomena in Electronic Systems, Orlando, FL (2008).

28. Wimax TI Library. [Online]. Available: http://www.ti.com/corp/docs/landing/wimax/index.htm.
29. Witte E.M., et al.: SDR baseband processing portability: A case study. In: Karlsruhe Workshop on Software Radios, Karlsruhe, Germany (2008).
30. Wurm P.: A digital LINC radio transmitter architecture for opportunistic radio. In: IEEE Veh. Technol. Conf., Barcelona, Spain (2009).

Green communications

Hanna Bogucka (✉), Jacques Palicot, and Christophe Moy

Since the United Nations General Assembly in December 1987, and its Resolution 42/187, sustainable development has become an issue and an aspiration of our civilization. The most often-quoted definition of sustainable development has been formulated by the Brundtland Commission [1] as the development that "meets the needs of the present without compromising the ability of future generations to meet their own needs."

European Union in particular is continuously making efforts in legislation, regulations and recommendations to all sectors of the industry and agriculture, as well as in promoting ecology, and ecological life-style to address anticipated future climate issue and sustainable development in general. Surprising as it may seem, Information and Communication Technology (ICT) domain should be particularly concerned. (As an example, power consumption and CO_2 emission data for cellular networks as of today are presented in Table 1) Currently, 3% of the world-wide energy is consumed by the ICT infrastructure (out of which 80% is consumed by the base stations, what amounts in approximately 5-10 million KW/h of energy each year) which causes about 2% of the world-wide CO_2 emissions. Note that this is comparable to the world-wide CO_2 emissions by airplanes or one quarter of the world-wide CO_2 emissions by cars. As the transmitted data volume increases approximately by a factor of 10 every 5 years, this corresponds to an increase of the associated energy consumption by approximately 16-20% per year. So energy consumption is doubled every 5 years, which may cause serious problems in terms of sustainable development.

Moreover, the Wireless World Research Forum (WWRF) envisions that 7 trillion wireless devices will serve 7 billion people by the year 2017, and thereby, people will always be connected to the Internet [3]. Therefore, lowering energy consumption of future wireless radio systems is demanding considerable attention at all the stages of the telecommunication systems [4]. The approach to lower the energy consumption and related CO_2 emission is known as Green Communications (GC). Green communications philosophy may apply to all stages of the ICT chain, from applications and infrastructure, to radio access points and equipments. Green Communication is a major challenge for next 10 to 20 years. Green communications indeed may be gen-

Benedetto S., Correia L.M., Luise M. (eds.): The Newcom++ Vision Book. Perspectives of Research on Wireless Communications in Europe.
DOI 10.1007/978-88-470-1983-6_9, © Springer-Verlag Italia 2012

Table 1. Power consumption and CO_2 emission of cellular networks (based on [2] NSN data)

	Power consumption of a single device	*Number of devices worldwide*	*Total power consumption*	*CO_2 emission per year*
Mobile terminal	0.1 W	≅ 3 billion	0.2–0.4 GW	≅ 1 ÷ 2 Mt
Base station	1 kW	≅ 3 million	4.5 GW	≅ 20 Mt
Infrastructure unit (CS and PS Cores, IP core, transport system)	10 kW	≅ 10 thousand	0.1 GW	≅ 0.5 Mt

eralized beyond climate matters, to the efficient utilization of the natural resources and to people's health. ICT development during next half of the century may be conditioned by GC considerations, for the ecological and health reasons, and it will also be critically assessed by the public opinion.

1 A holistic approach to wireless-communications sustainable development

A holistic approach to the wireless communication eco-sustainability is based on the idea that all wireless system and network components as well as all the infrastructure components should contribute to the overall sustainable development of the considered ICT sector, and that the properties of that sector related to the impact on the eco-system cannot be determined only by its component parts alone. (The general principle of holism was concisely summarized by Aristotle in the *Metaphysics*: "The whole is more than the sum of its parts.") In other words, the eco-sustainability of the wireless network is not only about just the network elements, rather it is about the whole network development which would take all the relations and interactions between the network components to achieve the wireless communication sector sustainable development.

Thus, the attention has to be given to multiple network components and their dependencies to come out with noticeable power savings of the communication network. As an example, in Table 2, some power reduction options at various communication system layers are summarized.

Apart from the typical wireless network methods, that add flexibility to the system layers to optimize the power consumption a great deal of opportunities occur in the network and infrastructure management to reduce typical power requirements. These management procedures can be envisioned as follows:

- auditing of the network power strategy to identify areas for efficiency gains, including alternative energy opportunities;
- consolidating options for energy-efficiency at the required operational performance;
- decommissioning of obsolete or under-used assets;

Table 2. Options for reducing the environmental impact of wireless networks (based in part on a comparison of the Alcatel-Lucent product generations [2])

	Example methods and measures	*Power savings*
Hardware platform	High-class, energy-efficient multicarrier power-amplifiers	30%
Algorithms/Software	Dynamic power savings, adaptive transmission techniques	30%
Network level	Smart antennas	40%
Infrastructure	Remote radio head	25%

- sharing assets to optimize, streamline and reduce network resources and operating expenses;
- advancing power-efficient cooling systems and solar panels;
- introducing sustainable solutions to extend the life of investments;
- remanufacturing, recycling.

Two major axes are to be distinguished within Green Communications. The first one concerns the telecommunication infrastructure (which includes the transmission over the wired core network, switching, network management and maintenance) and is mostly concerned with the energy-savings matters. Second is at the radio access technologies. Energy consumption is still of importance for wireless devices, their users and their environment, and one shall distinguish the energy consumption issues between the energy consumption for the signal processing, RF front-end processing and amplification, HW power supply, etc, which has an impact on the global energy-consumption and thus also the global warming, and energy radiation that will have a major impact on human health and so-called electromagnetic pollution. However, it seems that the later problem is related to the radiation power. This is because instantaneous radiation of high energy (high power) can be more harmful to the human health and for sure is the source of electromagnetic interference of various nature.

As mentioned above the concept of Green Communications results from the emerging global consciousness concerning the ecosystem. It is noticeable that notions such as consciousness, intelligence, cognition, sensing, so far closely related to the human nature, are being used for the description of artificial intelligence and for the functionalities of various decision-making and learning devices, including designs for modern wireless communication. Cognitive Radio (CR) is such an emerging concept of an intelligent communication design that can orient itself in the radio environment, create plans, decide and take actions [5]. We believe that consciousness is still related to a human being only, and it is an umbrella above the notion of cognition. It seems however that cognition alone of the CR devices can be used for making the Green Communications the reality at the radio access level of the ICT infrastructure. In fact, many researchers look at the CR as an enabling technology for reasonable utilization of radio resources, namely power consumption and available frequency band. Moreover in this context, "classical" understanding of green communications (i.e. power consumption) can be extended towards many other aspects of radio communications.

2 Green Communications for the climate issue

2.1 Flexible and cognitive radios for energy efficiency

Green Communications is mostly seen as a method to address the climate-changes issue by limiting the CO2 emission. It mostly consists then in considering usual studies in a new way. Let us consider the following classical topic of modern flexible radios, including CR which is the broadly-understood optimization of radio systems behaviour. But the goals for the optimization may be plural, including some rationality in utilizing available resources (bandwidth, power, ...) or fairness. It is usually the matter of the appropriate definition of the objective function in this optimization procedure. From a theoretical point of view, the gain in spectrum efficiency could be used to decrease the radiation level. However, from a practical point of view, most of the actors in the telecommunications domain have been using this gain to increase the bit rate with a constant power level instead of maintaining a constant bit rate with a lower power level. Therefore, a green approach of flexible and cognitive radios could give the priority to "political" and "sociological" matters, instead of financial considerations. The optimization procedure can be used to decrease the electromagnetic radiation level by sending the right signal in the right direction with the optimal power and only when it is necessary, i.e. a requested transmission rate is to be guaranteed. It is exactly what green communication is about. In this context, using all the information on the user's environment provided by sensors [6], CR can provide efficient solutions [7]. In order to achieve these expectations, research work [8] has to be done into the following domains:

- signal processing techniques (spectrum sensing, channel estimation, identification, positioning, adaptive transmission schemes, peak-to-average power ratio reduction schemes, beam forming, smart antennas, etc.) in order to sense the environment, and apply energy-optimized signal processing and transmission procedures;
- computer science (decision making [9–11], learning, prediction [11], etc.) to make the adequate decisions;
- hardware and software reconfigurability capabilities to modify and adapt the equipment behaviour and its power consumption [12].

Let us give an example on the energy consumption minimization by the use of adaptive coding. In [13], a terminal (or a base station) is considered, which is able to determine the transmission channel quality in real time, and adapt (also in real time) the coding rate to minimize transmitted power (and resulting Signal to Noise Ratio (SNR)), while keeping the Bit Error Rate (BER) sufficient for the guaranteed Quality of Service (QoS), as it is illustrated in Fig. 1 (see the horizontal arrow). This simple example is not new. In fact, several norms and recommendations already today include this type of processing. Adaptive modulation and coding for the energy efficiency has been addressed by the project WINDFLEX of the European Union 5th Framework Programme [14].

Similarly, a terminal could also decide to change other parameters of the physical layer with this objective in mind. For example, it could decrease the symbol rate if

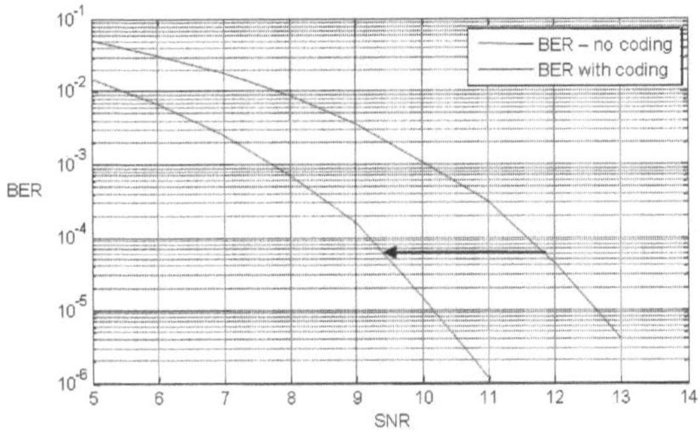

Fig. 1. Gain in performance versus power decrease

the reception duration is compatible with the required QoS. In [15], the degrees if freedom for energy savings in practical wireless transceivers are discussed, for both single- and multicarrier transmission. There, it has been shown that there exist a lot of opportunities in adaptive schemes to save the energy, e.g. in the application of the pilots energy, in the number of applied diversity branches, the constellation order, etc. The results of the simulations have been also provided showing a energy-gain when applying the considered adaptive schemes. The energy-efficiency objective can be extended to use other transmission parameters and gains such as the MIMO gain, beam forming gain, etc. To sum-up, current standards are dimensioned in order to support a "worst case" of utilization or in order to increase the spectral efficiency by maximization of the transmission rate. Energy-efficient radios should be operating in a "just enough case", thus optimizing the natural resources, which are power and spectrum, and reducing the side effects on health.

2.2 *Wireless communications to support smart power grids*

The concept of a smart grid to deliver electricity from suppliers to consumers using digital technology is relatively new, and aims at controlling appliances at consumers homes and premises to save energy. Additionally, it is supposed to reduce cost and increase reliability and transparency of both the suppliers and the consumers. It is promoted by many governments to deal with global warming and other issues, such as energetic safety.

A smart power grid will consist of devices that plug into an outlet and into which a customer may plug his appliance. These devices communicate and report to the electric companies multiple information concerning the time, amount and efficiency of using the energy. The idea is to optimize the charge of using the electricity for peak hours and low demand hours, what would result in coercing desirable consumers behavior to save energy. Apart from monitoring the energy consumption and optimization of the electricity demand, smart power grids are envisioned to handle energy efficiency and energetic safety through avoiding power overload, e.g. by selectively turning some of the devices off, as well as to remotely monitor and control used appliances.

For the idea of smart power grids to happen, the need of robust, scalable and reliable communications infrastructure has been identified. It is supposed to be based on cooperative communications in home-area networks connected through wide-area networks. The privacy and safety of this communication is also a major issue. In the near future, wireless telecommunication engineers and researchers will have to develop wireless communication technology to support the smart power grids development.

3 Green Communications for electromagnetic radiation level reduction

3.1 Green Communications for a better spectrum use

One of the challenges of the future wireless radio systems is to globally reduce the electromagnetic radiation levels to obtain a better coexistence of wireless systems and lower interference. Thus, any signal processing study investigating ways to increase the wireless networks capacity are contributing to Green Communications. Let us just take a few examples of the trendy topics: MIMO processing, near-Shannon limit channel-coding schemes, cooperative communications and relying, etc. All the signal processing research lead in this sense in the next 20 years are contributing thus to the Green Communication field.

Cognitive radio is also addressing the same idea while proposing solutions to optimize the spectrum resources. Cognitive radio is indeed often reduced to that particular point [5] despite more general definition of Joe Mitola in [17]. Network management aspects of CR in particular are bringing new perspectives in solving the capacity issue in terms of a better spectrum resource use. Cognitive Networks (CN) concept indeed aims at managing networks and their interconnections more intelligently by accessing and intelligently using the information available thanks to the networks, systems and terminals cognition. A terminal or a base station could recognize the spectrum occupancy and therefore decide which band (with the associated power) is the best from the point of view of electromagnetic radiation. A voice service for instance can be supported by a GSM call instead of using a 3G channel for that. This could be performed with the help of the standard recognition sensor. The concept of blind recognition [6] aims at blind identification of standards in use in the vicinity of the terminal. The following step would consist in reconfiguring the equipment, so that it could use the most appropriate standard according to service requirements, user needs and electromagnetic radiation efficiency.

Moreover, new adaptive spectrum sharing models should be designed, and cooperative communication techniques at all layers can be applied to increase the spectral and power efficiency, to enhance network coverage, and to reduce the outage probability in wireless networks. As a result, one may expect better spectrum usage in available bands, release of some other bands that should be reserved for other purposes, e.g. radiation detection of various nature, lower transmission power and thus, lower interference. Cognitive and cooperative communications with all their aspects of the radio environmental awareness and flexible and efficient usage of resources, are long term objectives, and thus, their related issues of Green Communications will concern many researchers for next 10 to 20 years.

3.2 Green Communications for spectrum pollution

In the sense of sustainable development, not only the energy consumption for CO_2 emissions is considered, but also the spectrum pollution with the increase of radiation related to communication systems. Spectrum may be considered as a natural and public resource which should be carefully used, shared world-wide, and economized. Spectrum indeed only exists if there is sufficient energy to generate the electromagnetic waves. Thus, spectrum savings is a question of capacity versus power spectral density of radio transmission.

Another aspect is to preserve radio bands for research and studies in natural sciences. This is particularly important for astronomy-related research fields. Astronomers complain more and more about electro-magnetic pollution. Reference [18] deals with the *destructive impact of civilization on current observational astronomy. Ironically, the technology that has made possible so many exciting astronomical discoveries is now jeopardizing the future of observational astronomy.* Astronomers need free bands so as to be able to make observations in radio-astronomy bands but also in millimeter-wavelength bands. To give just a few examples, solar and pulsars observations are made in the VHF and UHF bands (150 MHz and 410 MHz), spectrum lines in the L-band (Hydrogen at 21 cm), the Ka-band (water, Ammonia), etc.

The National Radio Astronomy Observatory of the US (NRAO) gives the following example in [19]: *radio-astronomy signals coming from the universe are millions of times weaker than signals used by human communication systems. Radio-astronomers would consider that a cellular phone on the Moon is a strong radio source.* Fig. 2 from NRAO shows how a radio transmission may pollute a radio-astronomy band. The blue band is reserved for radio astronomy measurements and the red level is the transmission level limit beyond which radio transmission is detrimental for radio-astronomy.

This problem in the next years will most probably stimulate more philosophical discussion on respecting other research domains. As stated in [20], *almost everyone can relate, first-hand, to the issue of light pollution. It is more difficult for people to relate to the astronomical windows in the electromagnetic spectrum.* As radio astronomers are often not understood, as it is being articulated within the scientific organizations such as URSI which are dealing with all scientific domains of radio sciences, that they express their concerns about radio communication development. *Green* approach to radio communication, through related concepts, e.g. cognitive radio, has the potential to change this situation thanks to a smarter use of the spectrum, and thus, reduction of the radio transmission pollution. Going towards this direction is a way to reassure radio-astronomers in the future, and to show that their call has been heard. The case of astronomy is given as an example here, but this could be gen-

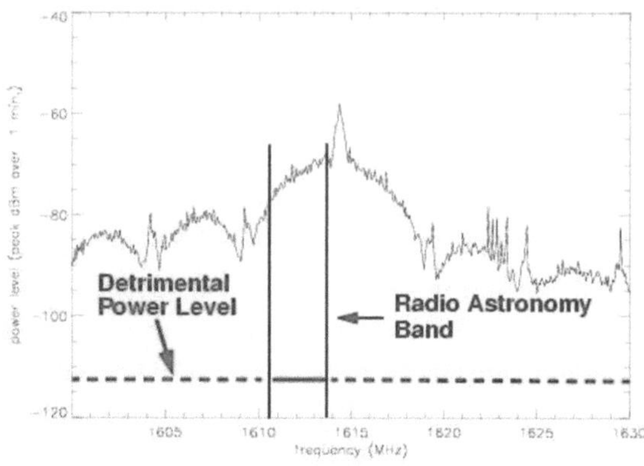

Fig. 2. Radio-astronomy band pollution by the radio-communication [19] (reproduced with permission of National Radio Astronomy Observatory)

eralized to any other domain, such as military bands or radar bands. Some military bands may be not used for transmission but for detection goals such as missile detection, or other. As a Green Communication topic, the co-existence with other users of the spectrum is a major challenge of the next 20 years of radio-communications development.

3.3 Green Communications for health

Another challenge of future wireless radio systems is to globally reduce the electromagnetic radiation levels in order to reduce human exposure to radiation [4]. The impact of future radio communications on the human health has been invoked in [21]. There, the main aspects of this issue have been considered, mainly related to the head exposure on electromagnetic fields and radiation. The radiation produced by an omni directional antenna close to a human body is divided into sectors in accordance with the distance from the obstacle [22]. Fig. 3 gives the illustration of the radiation pattern at the proximity of a head. There is a great reduction (> 30 dB) in the direction opposite to the main lobe. Therefore, a great part of the radiation energy is absorbed by the human body.

In a CR context, we may expect that a terminal knows where the head of the user is, thanks to the appropriate sensors [6]. With the help of smart antenna techniques or sectorial antenna technology, it will then be able to form the radiation diagram in order to reduce (and even cancel) the radiation emitted towards the user's head. This involves mobile positioning and real-time antenna diagram calculations which remains a major research challenge. The human exposure area is such an important issue that all possible solutions to mitigate electromagnetic radiation impact on human beings should be investigated within the next 20 years.

Another way of limiting persons exposure to radiation is illustrated in Fig. 4 [14]. The figure presents the situation, in which an indoor Wifi is present and connected

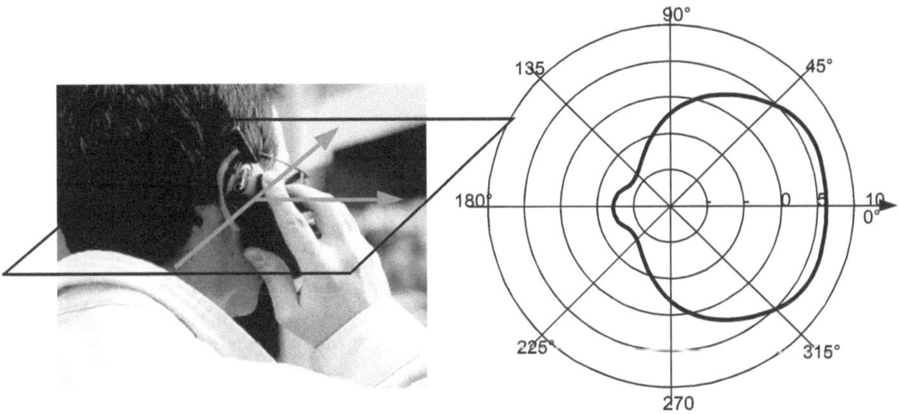

Fig. 3. Radiation pattern at the proximity of the head

to a cellular or broadband access network. Usually, in such a case it is preferable to choose the local connection. From the radiation level point of view it is always better to have a wired connection indoor and to have a roof connection with the cellular network. This is the kind of decision a modern radio equipment should be able to take, depending on the service asked by the user. The benefit here is double. On one hand, radiation power is limited inside the building, where people are living. On the other hand, a reduced-power transmission is possible to access the network as a directional antenna can be used on the roof to access the network.

Finally, another issue to be taken into account is the public opinion worrying about the multiplication of radio transmitters in the neighbourhood and inside living areas. People are more and more complaining about the multiplication of cellular transmitters near their home. In many countries, e.g. in France, a mood of suspicion about radio waves is noticeable, and more and more lawsuits are brought against the installation of new cellular base stations. Justice decisions are more and more taken

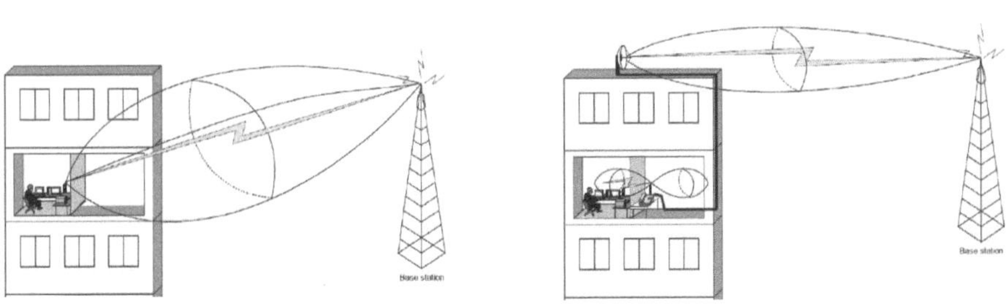

(a) (b)

Fig. 4. Wireless Internet access: (a) direct from terminal; (b) wired inside the building and then through a centralized wireless directional access point outside the building to decrease the transmission power

in favour of a complainant. First trials concerned school areas and the "principle de precaution" has been applied. But in mid-2009, a base station installation has been sentenced to be removed in a residential area, because of the proximity of the complainant home.

4 Radio equipment issues for Green Communications

The power-efficiency of the radio hardware equipment cannot be overestimated as its contribution to the mobile-terminal, base-station and overall-network power consumption and related CO_2 emission per year is noticeable (see Table 1). The silicon components optimization towards lower power consumption in Watts per Gigabit of the data processing is recognized as the continuously hot research topic. Moreover, handset system design is a continuing area of innovation (examples include hardware design to improve music playback time). Nevertheless, it has been recognized, that not enough attention is paid to the behavior of software applications taking into account an increasing number of applications relying on web services. In the next 5–10 years application developers and engineers should modify the applications taking into account many aspects of the end-user context (e.g. location, time-of-day) as well as battery levels.

5 Conclusions

European Union is taking efforts to address the climate issues including the ICT related issues and footprints on the environment. The European Commission concentration cluster "Radio Access & Spectrum" (RAS) aims to provide a platform for exchanges and concentration between FP6 and FP7 projects. Many of these projects address innovative radio transmission technologies to support future energy-efficient systems and networks. Objective 1.1 (The Network of the Future) of the EU 7th Framework Program, for the subsequent proposal calls defines objectives to support spectrally and energy-efficient networks, e.g. *Flexible and cognitive network management and operation frameworks* (under the *Future Internet Architectures and Network Technologies* objective), *Spectrum-efficient radio access to Future Networks* and *Converged infrastructures in support of Future Networks.* The effort to implement the idea of green communication is ongoing in Europe, but it is being promoted world-wide, what can be observed at multiple international conferences, plenary sessions and panels bringing together representatives from the scientific world, industry, operator companies and regulatory bodies.

It is of major importance for the sake of radio domain in the 21st century to bring progress to people and to their confidence, and not to make them fear of the radio evolution. Green Communications developments should provide this confidence as an ICT approach for finding solutions to the issues of CO_2 emission, energy consumption, resources utilization, electromagnetic pollution and heath issues. Undoubtedly, this can be considered as a challenge for the next 20 years.

References

1. United Nations, Report of the world commission on environment and development. General Assembly Resolution 42/187 (1987).
2. Panel Session Wireless future: Green and cooperative networking? International Symposium on Wireless Communication Systems, Siena, Italy (2009).
3. Wireless world research. [Online]. Available: http://www.wireless-world-research.org.
4. International Workshop on Green Wireless. [Online]. Available: http://www.cwc.oulu.fi/workshops/W-Green2008.pdf.
5. Haykin S.: Cognitive radio: Brain-empowered wireless communications. IEEE J. Select. Areas Commun. **23**(2), pp. 201–220 (2005).
6. Palicot J., Moy C., Hachemani R.: Multilayer sensors for the sensorial radio bubble. Physical Communication **2**, pp. 151–165 (2009).
7. Moy C., Bisiaux A., Paquelet S.: An ultra-wide band umbilical cord for cognitive radio systems. In: Proc. Personal, Indoor and Mobile Radio Commun., Berlin Germany (2005).
8. Moy C., Doyle L., Snada Y.: Cognitive radio: From equipment to networks. Annals of Telecommunications **64**(7-8) (2009).
9. Rieser C.J.: Biologically inspired cognitive radio engine model utilizing distributed genetic algorithms for secure and robust wireless communications and networking. Ph.D. dissertation, Virginia Tech (2004).
10. Colson N., Kountouris A., Wautier A., Husson L.: Cognitive decision making process supervising the radio dynamic reconfiguration. In: Proc. Cognitive Radio Oriented Wireless Networks and Commun., Singapore, Singapore (2008.)
11. Jouini W., Ernst D., Moy C., Palicot J.: Multi-armed bandit based policies for cognitive radio's decision making issues. In: Proc. Conf. Signal Circuits and Systems, Jerba, Tunisia (2009).
12. Thilakawardana D., Moessner K.: Traffic modelling and forecasting using genetic algorithms for next-generation cognitive radio applications. Annals of Telecommunications **64**(7-8), pp. 535–543 (2009).
13. Godard L., Moy C., Palicot J.: An executable meta-model of a hierarchical and distributed architecture management for the design of cognitive radio equipments. Annals of Telecommunications **64**(7-8), pp. 463–482 (2009).
14. Palicot J.: Cognitive radio: An enabling technology for the green radio communications concept. In: Proc. Int. Wireless Communications and Mobile Computing Conference, Leipzig, Germany (2009).
15. Polydoros A., et al.: WIND-FLEX: Developing a novel test-bed for exploring flexible radio concepts in an indoor environment. IEEE Commun. Mag. **41**(7), pp. 116–122 (2003).
16. Bogucka H., Conti A.: Degrees of freedom for energy savings in practical adaptive wireless systems. IEEE Commun. Mag., March/June 2011, to appear.
17. Mitola J.: Cognitive radio: An integrated agent architecture for software defined radio. Ph.D. dissertation, Royal Institute of Technology (KTH), Sweden (2000).
18. McNally D.: The vanishing universe. Cambridge Univ. Press, Cambridge, UK (1994).
19. Radio Astronomy and Interference. [Online]. Available: http://www.nrao.edu/index.php/learn/radioastronomy/interference.
20. Percy J., Preserving the astronomical windows by/for education and culture. ASP Conference Series **139** (1998).
21. Palicot J., Roland C.: On the use of cognitive radio for decreasing the electromagnetic radiations. In: Proc. URSI General Assembly, New Delhi, India (2005).
22. Koulouridis S., Nikita K.S.: Study of the coupling between human head and cellular phone helical antennas. IEEE Trans. Electromagnetic Compatibility **46**(1), pp. 62–70 (2004).

Conclusions

This book identifies medium-long term research tendencies/problems in the area of wireless communications.

The first paper contains a view on trends currently characterising the ICT world, with special focus on wireless technologies, based on the activities carried out within WPI.6.

Personal needs are addressed at the beginning, around various areas, i.e.: information and learning, environmental fitness, social interaction, working life, health and wellness, security and privacy, entertainment, and transaction. A brief description is done of each one, followed by a bridging to technologies, services and applications. This listing is intended to provide the rationale for the trends observed in ICT and corresponding technologies growth and evolution.

An overview of ICT trends follows via an original metaphor of the current ICT picture and biological DNA. Two meta-trends are identified, around personalisation and distribution, a discussion on these concepts being provided. Then, ten trends are identified: Ideal performance, Ubiquity, Flexibility, Complexity, Cognitivity, Opportunism, Cooperation, Security, Miniaturisation, Convergence. An analysis of each one is provided, which includes a mapping onto the two meta-trends.

Finally, cutting-edge standards and technologies in the ICT wireless context are addressed: LTE-A, WiMAX, MANETs and BANs, DVB-T2, DVB-SH, Digital Radio, Cognitive Radio, and Future Internet. For each one, a mapping onto ICT trends is provided, showing which trends are embodied by the technology and which personal needs can be satisfied in this manner. Along with this analysis, an estimate of the time frame characterising each of these standards is presented, together with a forecast the market impact of each technology, and a measure of the potential for scientific research underneath each standard, with the objective to guide NEWCOM++ researchers towards the most cutting-edge thematic.

The second paper addresses wireless networks, with emphasis on heterogeneity and opportunism. This paper provides some views on the possible evolution of future wireless networks, on some of the envisaged elements that will guide this evolution, and on the design challenges that are currently at stake.

Benedetto S., Correia L.M., Luise M. (eds.): The Newcom++ Vision Book. Perspectives of Research on Wireless Communications in Europe.
DOI 10.1007/978-88-470-1983-6_10,

The implications of network heterogeneity on how wireless networks will be operated and perceived by the users are presented, with the corresponding requirements for smart radio resource management strategies. The basic ingredients allowing to successfully coping with the increasing demand of traffic and bit are addressed: increase of the spectral efficiency, i.e., the number of bits/s that can be delivered per unit of spectrum; increase of the number of base stations to provide a service in a given area; and increasing the total available bandwidth. Within this area, the heterogeneity in networks and devices is also addressed, together with the multiplicity of data services with high bit rate requirements. Trends towards decentralisation and flat architectures, and perspectives on flexible spectrum management are shown as well.

Self-organising networks are included in the analysis, since it is expected that many mechanisms will be established to allow the self-configuration of a network with minimum human intervention. The features to be included in such an approach include: self-configuration, self-planning, self-optimisation, self-managing, and self-healing.

Opportunistic networks are also addressed, linking with Delay tolerant Networks (among others), vehicular communication networks are analysed, linking with Intelligent Transportation Systems, and the Future Internet is viewed from the Internet of Things and RFID perspectives.

The third paper also addresses wireless networks, but focusing on cognition and cooperation, and providing a view on efficient ways of setting networks.

When addressing Cognitive Radio Networks, several aspects of spectrum usage and management are discussed. A number of techniques to be developed for the implementation of efficient spectrum usage through cognitive radio networks are dealt with: spectrum sensing, spectrum management, spectrum mobility, and spectrum sharing mechanisms.

Cognitive Positioning is then addressed, in relation with cognitive radio, Multi Carrier systems being taken as an example. Signal-intrinsic capability for localisation is discussed, namely on the criteria to optimise the function of ranging, and on the characteristics of the signals.

A brief discussion on cooperative networks follows. Colocation of base stations and MIMO relay schemes are among the topics listed in the sections.

Finally, the concept of Docitive Radios & Networks is introduced, i.e., a novel framework on radios & networks that teach other radios & networks. These radios & networks are not (only) supposed to teach them the end-result, but rather elements of the methods to getting there. A taxonomy is presented, together with an example, a vision, and challenges.

The fourth paper presents the main facets of research challenges that lie ahead towards the goal of understanding the ultimate performance limits of networks, and of designing innovative techniques to approximate and even achieve them.

Various optimisation- and control theory driven techniques where put forward such as distributed optimisation and the max-weight control principle. Novel and disruptive approaches are described, such as network coding, and their potential is analysed. Furthermore, mathematical tools such as cooperative and non-cooperative game theory, learning theory and stochastic geometry will be needed in order to model and understand the spontaneous interactions of massively dense autonomic networks. This paper presents a holistic approach with techniques that are amenable to realisation in various classes of wireless networks, from sensor networks to infrastructure-based and infrastructure-less wireless networks.

Clearly, the way ahead is promising. Novel and disruptive approaches will need to be undertaken, oftentimes relying on cross-disciplinary techniques migrating from a wide range of disciplines and thrusts as described above.

The fifth paper discusses theoretical and practical problems on bandwidth- and energy-efficient radio access. The first part of the paper discusses on approaching the fundamental communication limits, by addressing them, and then, by showing some techniques that enable this approach, namely coding/decoding techniques, as well as adaptive modulation and coding.

Afterwards, iterative techniques in wireless receivers are presented, including the framework for iterative processing, applications on iterative receiver design, and low complexity decoding.

Finally, cognitive spectrum usage (always for the purpose of advancing spectral utilisation) with a special focus on applied game theory is discussed, including multicarrier techniques and distributed spectrum usage.

The sixth paper addresses on multimedia signal processing for wireless delivery, encompassing a discussion on the current situation, a review of enabling technologies, and a view into the global optimisation problem.

A discussion on the general trend is presented, including some of the current technologies, presenting some of the driving forces, and extracting general consequences for future work.

The review on the enabling technologies addresses 3-D representation (based on stereo vision and signal processing), screens and cameras, the network (e.g., mobile ad hoc networks and network coding, and possible research direction).

The global optimisation issue includes cross-layer optimisation, joint source-channel-network-protocol coding, and services and usage from a wireless multimedia processing viewpoint.

The seventh paper fits within recent trends in silicon technology and communication system demands, which exhibit a growing gap between application needs and what the technology can deliver. A key driver for the telecom industry is the mobile business. Mobility, which relies on battery operated handheld devices, provide stringent requirements on equipments in terms of processing power, power consumption and flexibility. At the same time the battery and silicon technology does not progress at the same pace. The emergence of new standards implementing ever more efficient air interfaces also put stringent constraints on the design time. Thus, the reuse of hardware building blocks and a proper methodology and tools are needed to evaluate hardware performance tradeoffs at the earliest stage. Flexible radio is a promising approach in this regard. However, a unified framework is still to be found to enable the co-design of communication functions and hardware platforms.

The eighth paper deals with "green" communications, a concept drawing from the universally shared concern about climate changes and aspiration toward a sustainable development guided by ecological considerations. European Union is taking efforts to address the climate issues including the ICT related issues and footprints on the environment. The European Commission concentration cluster "Radio Access & Spectrum" (RAS) aims to provide a platform for exchanges and concentration between FP6 and FP7 projects. Many of these projects address innovative radio transmission technologies to support future energy-efficient systems and networks. Objective 1.1 (The Network of the Future) of the EU 7th Framework Program, for the subsequent proposal calls defines objectives to support spectrally and energy-efficient networks, e.g. *Flexible and cognitive network management and operation frameworks* (under the *Future Internet Architectures and Network Technologies* objective), *Spectrum-efficient radio access to Future Networks* and *Converged infrastructures in support of Future Networks*. The effort to implement the idea of green communication is ongoing in Europe, but it is being promoted world-wide, what can be observed at multiple international conferences, plenary sessions and panels bringing together representatives from the scientific world, industry, operator companies and regulatory bodies.

It is of major importance for the sake of radio domain in the 21st century to bring progress to people and to their confidence, and not to make them fear of the radio evolution. Green Communications developments should provide this confidence as an ICT approach for finding solutions to the issues of CO2 emission, energy consumption, resources utilisation, electromagnetic pollution and heath issues. Undoubtedly, this can be considered as a challenge for the next 20 years.

Appendix

NEWCOM++ Millenium Problems

This appendix of the NEWCOM++ Vision Book contains a list of *Millenium Problems*, seminal problems in the area of wireless communication networks, characterized by being crucial and still unsolved. The problems have been identified by NEWCOM++ researchers and filtered by the editors of the Vision Book. All millennium problems are described using the same template, for the sake of conciseness and uniformity.

Benedetto S., Correia L.M., Luise M. (eds.): The Newcom++ Vision Book. Perspectives of Research on Wireless Communications in Europe.
DOI 10.1007/978-88-470-1983-6_11,

Table A.1. MP1. SoC and multi-tasking in future radios

Title MP1. SoC and multi-tasking in future radios

Author Dominique Noguet

Description

Processors architecture is at the verge of a radical evolution where multi-cores will become prominent. Because the processor frequency rate has pushed microelectronics technology to its limits, one major trend to fill the gap between processors' computational power and ever more demanding wireless applications is to consider multi-cores (eventually many cores) architecture. Mapping wireless baseband processing onto such architectures is a very complex task, first because of the sequential nature of many wireless algorithms, then because communication between processor may become the new bottleneck to real time processing. The issue to be solved touches both hardware architectural and software research area. Whereas the first field will have to solve which processor and communication architecture shall be selected, the second one will have to come up with new methodology and applications to map the functions onto the processor array. One approach, called virtualization may be one way forward, but its generalization to high parallelism factors is unsolved yet. The overhead implied must be analyzed carefully too.

Background and motivation

Over the 70s and 80s, general purpose computers have tried to avoid introducing multi-processor architectures. This was mostly due to the fact that programming (or designing efficient compilers) for such machines was a very difficult task. The so called CISC processors have pushed the limits by implementing several processing units in the same processor (ALUs, FPUs) and by designing the complex set of instructions that went along with them, which compilers could exploit. In the 80s the complexity of CISC processors had gone so high that new architectures needed to be invented. Fortunately, the emergence of RISC processors (followed by superscalar) has enabled to avoid the multi-core pitfall for a while, because these processors could have similar performance as CISC but at a lower clock speed. In the 90s, the processors' architecture was fully mature and the rate for higher frequency was the only way to enhance processors performance. These higher frequencies could be reached by the evolution of microelectronic technology rather than by architectural innovations. In the 00's silicon technology seemed to have reached some limits and going for higher frequencies did not make sense anymore because of power dissipation problems. Nowadays, we see the emergence of multi-core architectures in our PCs, for the parallel core paradigms is unavoidable at the moment. The same trend has reached processors for wireless systems, where computational demands have boomed because of the quest for higher throughput and higher spectrum efficiency. However, research on where and how MP-SoC architectures must be used in a communication system (system level, block level) has just started. The proper communication architecture between processors (eg. Network on Chip) is to be tackled too. Finally, considering power consumption issues in limited form factor battery operated devices is yet another challenge to consider.

References

[1] Jerraya A.A.: Long term trends for embedded system design. In Proc. Digital System Design, EUROMICRO Systems, Washington, DC (2004).
[2] International Technology Roadmap for Semiconductors, http://www.itrs.net
[3] Karam L.J., AlKamal I., Gatherer A., Frantz G.A., Anderson D.V., Evans B.L.: Trends in multicore DSP platforms. IEEE Signal Process. Mag. **26**(6), pp. 38–49 (2009).

Table A.2. MP2. Flexible radio design

Title MP2. Flexible radio design

Author Dominique Noguet

Description

The emergence of many different radio standards has shifted radio designers into the flexible radio paradigm. Flexibility is often understood as "more software". However, as far as real-time radio design for ever more demanding modulation and coding schemes is concerned, a pure software solution (referred to as software radio [1]) is a myth. Key limitations are: power consumption, real-time operation, silicon usage efficiency. Hence a clear cut between the tasks that need to be tackled in hardware and those in software is not yet optimally found. Of course, commercial solutions of flexible radio (from a user experience point of view) are available today. However, when looking into how these are implemented, it can be observed that the Velcro approach is still very present, leading to under optimized or closed solutions. This is particularly true in the analog front end section. After the hype of the software radio in the mid 90s, we are entering a more mature phase, where realistic approaches emerge. However, how to optimize such solutions from a design methodology viewpoint and from a hardware architecture perspective is still very open.

Background and motivation

J. Mitola introduced software radio in the mid 90s. This conceptual approach to modern radio was to consider very high frequency converters (ADC and DAC) attached to a powerful processor that could handle any radio modulation and coding scheme. This concept has never been implemented as such due to many limitations. Converters with very large band and high dynamic range do not exist and cutting edge converters have dramatic power consumption figures. High power DSPs have high power consumption characteristics and cannot cope with extremely demanding recent standards for which hardware implementation may be preferred (e.g., for channel or MIMO decoders). For this reason, hardware platforms are either based on parallel frozen designs which are switched from one to another to address the requested standard, or a so called hybrid platform where hardware accelerators are mixed with processors. Most of the time, available architectures are updates of older versions, therefore inheriting the same global philosophy. Thus a system level design methodology taking the best of processors, accelerators, banks of common operators and programmable logic is still to be found. This optimization is not only a research area for academics, but it is a necessary path to make future radio possible for the industry. It is also a means to support the cognitive radio paradigm.

References

[1] Mitola J.: The software radio architecture. IEEE Commun. Mag. **33**(5), pp. 26–38 (1995).

Table A.3. MP3. Power consumption aware radio design

Title MP3. Power consumption aware radio design

Author Dominique Noguet

Description

Power optimization touches a wide range of competences in different fields. Improving power density storage is one way to make battery operation longer. Today, Li-Ion batteries offer the best power storage density, but the new era of nanotechnology may lead to disruptive technologies bringing outstanding performance.
At an architectural level, power optimization techniques can be exploited, like VDD hopping, frequency scaling, partial idling or clock gating. However, these techniques are at the very edge of what technology and design kits offer and new methodology must be found to enable power optimization throughout design time, by suggesting the inference of such 'tricks' to the designer.
Finally, microelectronic technology or nano technology will probably lead to significant breakthroughs. Because classical CMOS is reaching its limits, decreasing power consumption at technology level will request new transistors. Silicon on Insulator (SOI) is probably a good option to significantly improve power consumption while preserving current design methodologies. In a longer term, several nano technologies will bring electronic design to a new era, both in performance and in the way electronic components are modeled.

Background and motivation

Power consumption has become one of the key constraints for radio design optimization. This originates in several factors. The first one concerns handheld devices which are, most of the time, battery operated. As a matter of fact, the evolution of energy storage density has not evolved, by far, at the same pace as processing power and the power consumption that comes with it. This gives the designer the choice to either increase the device form factor significantly or to reduce the battery operation time. From a users' point of view none of these options are bearable. The last degree of freedom is to optimize power consumption at design time.
The second reason behind power consumption optimized design is the fact that power density in the chipset is coming so high that it could make either the chip to break, or the device to become too hot to hold, or the use of active cooling necessary. Of course, none of these options are satisfying.
Finally, the hype for green radio has made operators concerned about reducing their energy footprint. Besides, it has been observed that cellular networks operator power bills have raised the OPEX to unbearable limits. Therefore, although active cooling, reasonable form factor, and power to the plug can be provided for base stations, there are strong benefits of power consumption aware radio design at the base station side (and the core network) as well. Some initiatives such as the EU FP7 EARTH project or the worldwide Greentouch initiative are tackling this issue. It turns out though that optimizing for power leads to new technological solutions that need to be investigated further.

References

[1] Javey A., Kong J. (Eds.): Carbon Nanotube Electronics. Springer Science+Business Media, New York, NY (2009).
[2] Belleville M., Faynot O.: Evolution of deep submicron bulk and SOI technologies in low power electronics design. In: Piguet C. (Ed.): Low-Power Electronics Design. CRC Press, Boca Raton, FL (2004).
[3] MEDEA+ Scientific Comm. Work. Group, Towards and beyond 2015: Technology, devices, circuits and systems. Tech. Rep. (2007). [Online]. Available: ttp://www.medeaplus.org/web/about/scientific.php?request_temp=2015

Table A.4. MP4. Variation aware architectures for digital communications

Title MP4. Variation aware architectures for digital communications

Author Guido Masera

Description

Variations in operating conditions, process variations and soft errors are critical sources of non-idealities in modern digital circuits and their effect on system reliability is expected to rapidly grow in the next generations of technology. The traditional approach to reliability is based on over-conservative design, able to guarantee correct operation even in the worst-case conditions; however, in this approach, improved reliability is paid in terms of higher power dissipation and/or lower clock frequency. A different design approach can be followed to achieve at the same time power/speed performance and reliability: less conservative design constraints are imposed and error resilient methods are incorporated in some form. Generic circuit level techniques are available to mitigate the effects of hardware errors by means of redundancy, but this solution tends to be very expensive in terms of required area and dissipated energy. In several applications, such as digital communications, some amount of transient errors can be tolerated and compensated at the algorithm and architecture levels. For example, in iterative channel decoders, additional decoding effort (e.g. iterations) could be used to also correct limited amounts of transient errors caused by variations. The investigation of new algorithms and architectures capable of correcting transient errors is an important topic in the next few years and it will make possible the implementation of circuits more efficient than those with traditional worst-case approaches.

Background and motivation

Process variations are enhanced by aggressive scaling of CMOS technology and introduce random or systematic changes among dies or even within a single chip. Random dopant fluctuations, channel length variation and line edge roughness are the most important sources of process variations that make the circuit performance unpredictable and limit circuit yield and reliability. Device variations also depend heavily on applied voltage: scaled supply voltages help reducing the power dissipation, but increase the effect of process variations on circuit delay. As a consequence of process variations, energy efficiency and circuit reliability are contrasting design objectives and achieving good trade-offs between them has become a very critical challenge in nanometer design. Therefore proper solutions must be explored to mitigate the effects of variability in current and next technology generations. Several circuit techniques have been proposed to combat negative effects of process variations and other sources of transient errors. For example self-calibrating solutions are known, which measure variability and exploit knobs like body bias or supply voltage in order to control circuit performance. In general, these approaches suffer from excessive design complexity and inaccuracy in modeled variations. A different already explored idea comes from the area of fault tolerance: methods like duplication or triplication of circuit components enable error detection or correction. However the prohibitive area and power overhead associated to these solutions limit their use to the domain of safety-critical systems. Error tolerance can also be introduced at the architecture and algorithm level. For example, algorithmic noise tolerance is a known technique exploiting the robustness against occurred errors in several DSP algorithms to reduce power dissipation by means of voltage over-scaling: the same approach can also be adopted to combat errors caused by process variations. In this case, the largest part of the system under development is seen as a combination of unreliable components that introduce errors in their computation; additionally, a reduced number of fully reliable components are allocated to handle errors. A huge design space is available to implement error resiliency: additional reliable components may be exploited to support error detection and correction, at least in critical parts of the systems; alternatively, in stochastic applications, the algorithm itself, and the corresponding implementation architecture,

Table A.4. continued

may be modified to compensate for occurred errors; finally, methods have also been proposed to formulate applications as stochastic optimization problems, where reliable results are obtained from unreliable computation.

References

[1] International Technology Roadmap for Semiconductors, http://www.itrs.net
[2] Edge R., Shanbhag N.R.: Energy-efficient signal processing via algorithmic noise tolerance. In: Proc. ACM/IEEE Int. Symp. Low Power Electronics and Design, San Diego, CA (1999).
[3] Leem L., Cho H., Bau J., Jacobson Q.A., Mitra S.: ERSA: Error resilient system architecture for probabilistic applications. In: Proc. Design, Automation & Test in Europe Conf. (DATE), Dresden, Germany (2010).

Table A.5. MP5. High level design approaches for complex systems

Title MP5. High level design approaches for complex systems

Author Christophe Moy

Description

Electronic systems are becoming more and more complex as they turn more and more flexible. Such flexible systems' design is very challenging as it combines all the hardest points of multiprocessing, HW/SW co-design, flexible design with many use-case scenarios that cannot be 100% tested at design time. Higher level design approaches than current ones are required to make all actors of the design and development, whatever their background (HW engineers, SW engineers, system engineers, managers, etc.) understand and work together, based on models of the other developers' sub-parts. This particularly concerns the radio design of future cognitive radio systems and equipments which are a specific case of complex systems. A modeling perspective integrating all design steps, from highest levels of cognitive radio equipments specification down to the implementation, is a very challenging perspective. If we consider that a cognitive radio aims at optimizing the radio operation at any level (as stated in Mitola's first approach of cognitive radio [1]), power consumption and electro-magnetic radiations mitigation can be considered as an optimization target as spectrum use. Cognitive radio is then foreseen as an enabling technology for sustainability and green radio [2], thanks to a system approach, implying cognitive management architectures embedded in the equipment in relation with the management architectures of other entities of a cognitive network. This will help the acceptation by the public opinion of the proliferation of radio systems in everyday life.

Background and motivation

Electronic systems enter a new area of auto-adaptiveness. All future systems are foreseen to be auto-adaptive, in every field. Let us just give some examples: autonomous robots, autonomous transportation, auto-healing systems, etc. The common ground of all these systems is that they have to combine all three capabilities of intelligent systems: sense their environment, decide to adapt, and change their operation. Designing such complex systems is a challenge of the 21st century as the range of all possible scenarios of activation is becoming a very wide space. It consists in being able to describe in an integrated design flow the expected use-cases to be supported by the systems, derive the application operations, and finally specify hardware platform able to run the

Table A.5. continued

system, while offering facilities to balance the cost, the power consumption, the price, etc. Such a design approach does not exist today. It consists in merging many sub-design approaches and integrating them in a higher level global design perspective. That is the reason why we consider that high-level design is a key for future complex systems design. Any approach (tool, language, SW architecture, model) enabling to describe HW architectures, algorithms or application scenario at some abstraction level is worth.

References

[1] Mitola J.: Cognitive radio: An integrated agent architecture for software defined radio. Ph.D. dissertation, Royal Institute of Technology (KTH), Sweden (2000).

[2] Palicot J.: Cognitive radio: An enabling technology for the green radio communications concept. In: Proc. Int. Wireless Communications and Mobile Computing Conference, Leipzig, Germany (2009).

Table A.6. MP6. Cognition and Docition in distributed networks

Title MP6. Cognition and Docition in distributed networks

Author Lorenza Giupponi and Mischa Dohler

Description

There has been lately a clear trend towards decentralization of network architectures and radio resource management functions; IEEE P1900.4 standardization efforts and the femtocell paradigm are two remarkable examples. In this kind of networks, autonomous management and self-organization will provide significant benefits in terms of OPEX (Operational Expenses), ease of network management and reduced signaling burden over the backhaul. Cognition, referred to as the process involved in learning, gaining knowledge and comprehension, including thinking, knowing, remembering, judging, and problem solving, promises numerous advantages in this context over static or opportunistic approaches. However, the implications of learning in decentralized settings are still not fully understood even by the machine learning community experts and constitute an ongoing research line referred to as "multi-agent learning". The underlying interactive learning processes of the different nodes generate oscillating behaviors, which not always reach an equilibrium. The dynamics of learning may be long and complex in terms of required operations and memory, with complexity increasing with an increasing observation space. Other drawbacks are the slow rate of convergence and poor scalability, which essentially prevent these techniques from being used in dynamic large-scale decentralized systems. This problem calls for the definition of new learning paradigms capable of improving the learning process in terms of precision and speed.

Background and motivation

Learning in distributed systems is a quite unexplored field of research in the context of wireless communications, where radio resources have been traditionally managed by means of centralized approaches, and where the channel is affected by phenomena such as user mobility, lognormal shadowing, fast fading, multiuser scheduling, and most importantly, by radio resource management decisions of the other learning wireless nodes. In such a variable wireless environment, the main assumption of machine learning literature, i.e., stationarity of the surrounding environment, does not hold anymore. As a result, learning algorithms may be able to reach their objectives,

Table A.6. continued

but with a non satisfactory precision and only after very long and expensive training phases. This is where the docitive paradigm, introduced in the present deliverable, perfectly fits. In machine learning literature, some work has been realized in order to understand the benefits of knowledge transfer between different learning tasks, however, the implication of this in a wireless setting is still to be analyzed.

References

[1] Giupponi L., Galindo A., Blasco P., Dohler M.: Docitive networks – An emerging paradigm for dynamic spectrum management. IEEE Wireless Commun. Mag. **17**(4), pp. 47–54 (2010).
[2] Nie J., Haykin S.: A Q-learning-based dynamic channel assignment technique for mobile communication systems. IEEE Trans. Veh. Technol. **48**(5), pp. 1676–1687 (1999).

Table A.7. MP7. The impact of channel state uncertainty on MIMO multi-user precoding: Theory and practice

Title MP7. The impact of channel state uncertainty on MIMO multi-user precoding: Theory and practice

Author Shlomo Shamai (Shitz)

Description

While the impact of channel state information (CSI) availability and accuracy on MIMO point-to-point communications is reasonably well understood, this is not the case in MIMO multi-user wireless networks. The MIMO broadcast channel (downlink) is a classic example. While with full channel state information, the capacity region is known, and achieved by sophisticated 'dirty-paper' precoding, much less is known when CSI is partially available. Moreover, in the symmetric cases the lack of CSI causes the total collapse of the MIMO broadcast advantage (the multiplexing gain (degree-of-freedom (DoF)) reduces to 1, in case the downlink users are equipped with a single antenna), thus it is clear that this model exhibits sensitivity to the amount of available CSI.

Background and motivation

The theoretical understanding of these issues is incomplete, and even the fading scalar broadcast channel problem with no CSI at transmitter is still open. Practical needs of future wireless MIMO settings call for further research that will address:

- DoF in the presence of partial CSI;
- Limits of linear processing with partial CSI;
- Uplink/dowlink duality concepts with partial CSI;
- Aspects of interference alignment as a precoding method in partial CSI regimes;
- The impact of delayed CSI and retrospective based interference alignment and staggered fading models.

The deeper theoretical understanding will shed light into the practical precoding methods to be employed, which will deviate from classical precoding (either linear as zero-forcing, or non-linear as vector quantization). The preliminary indications demonstrate that there is room for robust interference alignment precoding, but how and what to do demands fundamental research on this aspect.

Table A.8. MP8. Impact of non-conventional traffic generated by M2M devices on cellular

Title MP8. Impact of non-conventional traffic generated by M2M devices on cellular

Author Raymond Knopp

Description

Machine-to-machine communications (M2M) is a very active area under discussion in 3GPP for integration within the LTE/LTE-Advanced framework. This type of application poses many interesting problems with respect to traffic modeling and the assessment of its impact on radio system design, in particular co-existence with conventional traffic sources. From a radio-access network perspective a proper understanding of M2M traffic is fundamental to the dimensioning of the underlying PHY and MAC procedures and in particular, adaptive coding and modulation and HARQ, multiple-access (MAC layer scheduling), design of short-length block codes, management/design of signalling overhead, joint source-channel coding. Below we motivate the study of this topic from the radio-access network perspective. Note that this can also be done in a more general networking perspective to cover core network aspects.

Background and motivation

One of the most challenging problems in future cellular networks is the co-habitation of M2M traffic with conventional user traffic. This is motivated by the likelihood of a rapid increase in the number of machines connected to cellular infrastructure in the coming decades (50 million machines out of 50 billion in 2008 connected to cellular networks!) It is clear that well before (i.e., now) the number of machines connected to cellular infrastructure starts to equal that of humans, the constraints imposed on the underlying air interface regarding latency and traffic scheduling policies will have to be considered in the standardization of new systems, starting with LTE-Advanced. This is already beginning within 3GPP RAN [1]. Fundamentally speaking, study of this type of traffic concerns the understanding of feedback signalling overhead and channel-code complexity (i.e. block-length) in the access stratum. The majority of wireless systems, including LTE, are designed for a continuous flow of information, at least in terms of the time-scales needed to send several IP packets (often large for user-plane data) containing information and such overhead is manageable. In some evolving M2M application scenarios (e.g. remote "smart" metering) packets are short and small in number and extremely low duty-cycle, while the number of these devices will be staggering, the combination of which from a system throughput perspective represents a vanishing data rate. In such low spectral-efficiency applications (seen by the application not the aggregate spectral efficiency), the signalling overhead latency translates directly into energy-loss, due to the fact that the whole embedded system is the sensing device is powered-up during the synchronization/training procedure prior to sending/receiving a short packet. While for a particular M2M or sensing node this energy-loss can be negligible, the aggregate cost due to the unbounded number of nodes could prove to be significant from a network standpoint. In addition, in the event that the number of sensing nodes connected to civilian infrastructure increases, the overhead factor for these services will significantly lower aggregate spectral efficiency of the network. This clearly calls for optimization of low-layer procedures and protocols with respect to overhead/latency for M2M applications. Another important aspect is the design of low-layer signalling which allows for extremely short acquisition times for event-driven traffic and switching between idle and active communication states.

Yet another key issue is that the type of information that M2M devices will generate/receive is often analog in nature at least at the source. Future research should revisit the classical problem of energy-efficient transmission of analog samples with minimal distortion at the receiving end, in particular from a network (i.e., multiple-sensor) perspective. They key issue with M2M traffic is that neither ergodic source sequences nor ergodic channel sequences can be exploited

Table A.8. continued

and that a feedback link (in the case of a sensor network on top of an LTE radio-access network for instance) should be considered. Furthermore, for analog information, the primary performance indicator is not high-throughput but reconstruction error as a function of the transmission energy. The multiple-sensor case is crucial since spatial-correlation between the sensing nodes must be taken into account in the design of the access-protocol. This correlation is reflected either in the sensed values themselves (i.e. sensing of a random field) or in the temporal arrival processes due to a common event detected by a set of sensors.

References

[1] 3GPP 37-868, Technical Specification Group Radio Access Network, "Study on RAN improvements for machine-type communications (Release 10)," Tech. Rep., July 2010.

Table A.9. MP9. Compact but accurate performance-prediction models for high-order MIMO with Maximum-Likelihood reception

Title MP9. Compact but accurate performance-prediction models for high-order MIMO with Maximum-Likelihood reception

Author Andreas Polydoros and Ioannis Dagres

Description

The AMC design can be viewed as a constrained-optimization problem with properly-defined cost functions and constraints so as to accommodate specific system scenarios. In order to perform this optimization, a simple-to-use (thus, of reduced complexity) yet sufficiently accurate link-performance-prediction functional model is required. This parameterizes the transmitted signal with enough granularity for any given channel and interference conditions. Thus, the difficulty in designing good AMC algorithms lies mainly in the proper (i.e., simple but accurate) construction of the overall optimization model, which can then be solved by various optimization techniques. Thus, the most challenging part is to find proper analytic models for link-level performance prediction for any complex system under consideration. Future systems will involve higher-order MIMO (in order to maximize link utilization) with corresponding ML reception for best performance. In these scenarios, reduced-complexity link-performance-prediction models have proven very difficult to derive, even for the simpler case of uncoded performance [1].

Background and motivation

With linear receivers the overall MIMO channel translates to an equivalent SISO; thus, the problem of choosing the optimal adaptive mode is simplified, allowing the practical use of AMC techniques for the equivalent SISO channel. Thus, linear MIMO receivers do not lead to any fundamental change in the AMC design procedure when the MIMO order increases, other than expanding the allowable Tx modes. With ML reception, things get more complicated. Conceptually speaking, Mutual-Information (MI)-based Effective SINR Mapping (ESNR) techniques achieve a similar mapping to an equivalent SISO model even for ML decoding algorithms. In practice, however, the model loses its simplicity even for a 2x2 system, making it useful only for off-line performance prediction and not for real-time parameter adaptation [2]. Various parameter-estimation errors, other non-linearities, or practical but sub-optimal ML receivers complicate the situation even more. Thus, developing proper, analytic, compact link-level performance prediction models amenable

Table A.9. continued

to run-time optimization, along with an elaborate trade-off analysis between performance-prediction accuracy on the one hand and complexity on the other is a very challenging part of the overall AMC design procedure.

References

[1] Kharrat-Kammoun F., Fontenelle S., Boutros J.: Accurate approximation of QAM error probability on quasi-static MIMO channels and its application to adaptive modulation. IEEE Trans. Information Theory **53**(3), pp. 1151–1160 (2007).

[2] Sayana K., Zhuang J., Stewart K.: Short term link performance modeling for ML receivers with mutual information per bit metrics. In: Proc. Global Telecommun. Conf., New Orleans, LA (2008).

Table A.10. MP10. A global Design of Wireless networks: from source to channel, including networking aspects

Title MP10. A global Design of Wireless networks: from source to channel, including networking aspects

Author Pierre Duhamel

Description

Scrutinizing the actual organization of wireless networks already demonstrates that some similar tasks are done at several of the separated layers. Error correction can be done at many places, as well as error detection. Joint decoding opened the way to a full use of all these redundancies, but by no means are we able to perform the best possible redundancy allocation for optimal performance.

Background and motivation

In a similar way, designing headers which contain a lot of redundancy, necessary for networking purposes, even if combined with header compression is certainly not the optimal procedure (even if the "separation theorem" was holding in the wireless context, we are not in the classical paradigm of source-channel separation, since we are designing the source...). Several such considerations tend to show that research should be addressing a global design of the network, from the source to the channel, including the protocol, where all aspects should be adapted to each other, even if some separation still has to be maintained (at what level?) in order to maintain the universality of networks.

Table A.11. MP11. Managed MANETs

Title MP11. Managed MANETs

Author Pierre Duhamel

Description

Actual research on wireless systems seems to go from one extreme to the other one: Actual wireless networks are totally centralized (i.e., all communications go through the base stations) and many studies are undertaken for self-organized networks (MANETs), where all users are "relaying" the communications of the other ones. Very little attention is paid to intermediate situations where the base stations could act as "conductors" without taking all communications in charge.

Background and motivation

Even if MANETs can ever be used (since they scale with difficulties), they will unlikely be the preferred solutions of network operators, since they will hardly make profit from such situations. In this case, even small scale self-organizing networks used to extend coverage or capacity of cells will hardly be pushed by network owners. This situation has already been met for Voice over IP, which is still not allowed in many cases, even if the technology is mature for a long time.
On the other hand, even when the base station plays an important role to organize the extension of the network, the operator will still manage the whole system, and will more likely accept such solutions. Such a situation has almost never been studied not even from a theoretical point of view. Such basic questions as what capacity improvement can be brought from such a situation, the scaling capacities, and even more the corresponding protocols are unknown.

Table A.12. MP12. Energy-efficiency of spectrum-sensing methods

Title MP12. Energy-efficiency of spectrum-sensing methods

Author Hanna Bogucka

Description

A lot of future novel transmission technologies, including cognitive radio, will be based on the assumption of the environmental awareness of the terminals. Despite some relatively stable, long-term unchanging information that might be available in some specifically designed data-bases (e.g. geolocation data-base of TV White Spaces), secondary-users will have to detect available spectrum, as well as primary-users transmissions appearing for short-term periods. This radio-spectrum awareness will require permanent sensing of possible spectrum opportunities as well as unexpected spectrum usage by the primary users. This sensing will be required from mobile devices, due to the fact that base stations, even if existing in the considered network, may not be able to detect local-range transmission in the mobile devices' neighborhood. However, permanent local sensing will consume a lot of the energy of the terminal.

Background and motivation

The sensing problem in unsynchronized networks with Dynamic Spectrum Access, becomes challenging as the secondary users of a cognitive radio network are not aware of exactly when the

Table A.12. continued

considered cases, the secondary users must sense the spectrum continuously and assure the priority to the primary-users transmission. Although the secondary users can benefit the most from continuous spectrum-sensing in terms of catching all spectrum opportunities, the continuous spectrum-sensing requires the secondary-users devices to always stay in the active state, and results in huge power consumption comparable to the transmit state [1]. This power-consumption related to continuous sensing is the major energy cost in the battery-driven mobile device. Sensing with a fixed sensing period may partially solve this problem. However, under the fixed sensing scheme, there may be undiscovered spectrum opportunities as well as unwanted interference caused to primary users transmission, that has not been detected. To overcome the energy-inefficiency problems related to sensing in the cognitive radio networks, some spectrum sensing have been proposed, which aim at the application of the adaptive sensing-period and proper sensing policies [1, 2]. They are however still not mature and do not result in significant energy-reduction.
Collaborative spectrum sensing has been also studied, not only to increase the detection probability or to minimize the total error rate but recently also for practical purpose of using energy-constrained sensor network for spectrum sensing in cognitive radio networks. In [3], the use of fewer sensors performing in a shorter sensing interval is discussed with the aim of preserving as much energy as possible while still satisfying the requirement for spectrum detection accuracy.

References

[1] Su H., Zhang X.: Power-efficient periodic spectrum sensing for cognitive MAC in dynamic spectrum access networks. In: Proc. IEEE Wireless Commun. Networking Conf. (WCNC), Sydney, Australia (2010).
[2] Su H., Zhang X.: Energy-efficient spectrum sensing for cognitive radio networks. In: Proc. IEEE Int. Conf. Commun. (ICC), Cape Town, South Africa (2010).
[3] Ngoc Pham H., Zhang Y., Engelstad P.E., Skeie T., Eliassen F.: Energy minimization approach for optimal cooperative spectrum sensing in sensor-aided cognitive radio networks. In: Proc. Wireless Internet Conference (WICON), Singapore, Singapore (2010).

Table A.13. MP13. Energy-optimized, scalable and parallel signal processing in wireless communication devices

Title MP13. Energy-optimized, scalable and parallel signal processing in wireless communication devices

Author Hanna Bogucka

Description

Future radio communications will aim at more and more content-heavy transmission (multimedia, augmented reality, etc.), based on reconfigurable radio platforms in order to exploit available radio interfaces in a flexible manner. This in turn will require significant processing power and flexible platforms that are usually considered as power-inefficient. In order to enable low-energy reconfigurable radio implementation suited for handheld multimedia terminals a two-step holistic approach is advocated combining efficient design with efficient operation. In both steps the ability of splitting the computational jobs into parallel tasks becomes a challenge, particularly given that the usual thinking on signal processing in communications presents subsequent operations performed one after another.

Table A.13. continued

Background and motivation

It has been recognized that there exists a gap between the power-requirements of content-heavy multimedia transmission and the power-availability in battery-powered handheld devices. Most of the researches dealing with flexible, reconfigurable platforms focus on reducing this gap by the means of more efficient processors. Several ASIPs (Application Specific Instruction Processors) and multiprocessor architectures have been proposed with power consumption acceptable for hand-held devices [1]. However, they are still far from being able to tackle signal processing and transmission of the latest MIMO-OFDM systems. Clearly, further innovations are still desired.
Besides the existing effort on processors and compilers, more research is required at a system level, particularly, platform-aware explorations and research on the baseband signal processing. With carefully optimized baseband signal processing implementations, the cost of the structure complexity is mostly an increased code size, which does not translate to significantly increased power consumption of the base-band architectures [2].
The approach presented in [1] is to partition the baseband signal processing implementation into a data plane and a control plane. The data plane consists of scalable baseband processing implementations, which allow a flexible tradeoff between the energy efficiency and the quality of processing. The control plane sets the knobs of the scalability to dynamically reduce the energy consumption [1]. For the data plane, highly parallel VLIW (Very Long Instruction Word) DSPs (Digital Signal Processors) are being studied to handle in real-time all the necessary signal processing of a wireless system transceiver (e.g., in [1] the implementation of WiMAX is studied using the abovementioned DSPs). However, in this case, the processes have to be efficiently mapped on parallel programmable architectures, which becomes a challenge, if the required degree of parallelism is high.

References

[1] Li M., Novo D., Bougard B., Desset C., Dejonghe A., Van Der Perre L., Catthoor F.: Energy aware signal processing for software defined radio baseband implementation. J. Signal Process. Syst. **63**(1), pp. 13–25 (2011).
[2] Van Der Perre L., Craninckx J., Dejonghe A., Green Software Defined Radios. Springer Science+Business Media, New York, NY (2009).

Table A.14. MP14. Activity detection, battery charge/discharge detection

Title MP14. Activity detection, battery charge/discharge detection

Author Hanna Bogucka

Description

Rechargeable cells in mobile devices have been tested and it has been observed that the charge recovery phenomenon is present in most rechargeable cells. Moreover, bursty traffic actually benefits from charge recovery, as cell has time to recover charge between bursts. To reduce energy consumption in wireless devices, it has been proposed that a charge state threshold should be selected (somehow based on the user's activity-characteristic). When the threshold is reached, discharge requests could be buffered until sufficient charge recovery takes place. Assuming a threshold can be determined, the question arises whether it is possible to realize algorithms for power savings that utilize this abovementioned phenomenon. Even when charge recovery is not possible, reducing rate of discharge is also desirable.

Table A.14. continued

Background and motivation

Today's cell phones support a variety of services beyond voice telephony, and battery life optimization is now being considered with respect to services with multimedia content. Both signal processing at the mobile and high-rate transmission is energy-consuming and calls for optimization. For the internet usage it is a less constrained problem than optimization for voice telephony or audio and video playback, because almost all popular internet services do not require a guaranteed level of quality-of-service (e.g. VoIP). As a result, power optimization in the web transport mechanisms is an attractive area of focus for internet power optimization, and has been considered for mobile proxy servers [1]. This is because a mobile can operate in a disconnected mode simply to preserve battery power. Overall, mobile web proxies have been recognized for their potential benefits in both extending battery life and also reducing the necessary over-the-air throughput for mobile web services. It has been reported that big-players, like Qualcomm are currently working on these issues [1].
It has been observed that the charge recovery phenomenon is present in most rechargeable cells. Moreover, it has been recognized that a battery in a mobile phone can recover charge between bursts of traffic. The challenge, however, is in determining and predicting the traffic bursts, particularly in the internet usage scenarios described above. It has been proposed that a charge state threshold should be selected (somehow based on the user's activity-characteristic). When the threshold is reached, discharge requests could be buffered until sufficient charge recovery takes place. Another challenge is in the proper method to determine this threshold and in finding and realizing algorithms for power savings that utilize this abovementioned phenomenon.

References

[1] Mandyam G.D.: Improving battery life for wireless web services through the use of a mobile proxy. In: Proc. IEEE Personal, Indoor and Mobile Commun. (PIMRC), Istanbul, Turkey (2010).

Table A.15. MP15. Bringing self-organization capabilities to networks

Title MP15. Bringing self-organisation capabilities to networks

Author Hanna Bogucka

Description

A Self-Organizing Network (SON) is defined as a communication network which supports self-x functionalities, enabling the automation of operational tasks, and thus minimizing human intervention. Self-organisation functionalities should be able not only to reduce the manual effort involved in network management, but also to enhance the performance of the wireless network, given the complexity associated with the large amount of parameters to be tuned. Due to its ambitious objectives, the practical realization of SON is seen as challenging and it can be anticipated that SON will continue as a hot research topic in coming years requiring further research efforts. Some of the aspects that must be properly covered include: to find an appropriate trade-off between the performance gains and the implementation complexity (e.g. required signalling and measurements, computing resources), to ensure the convergence to a stable solution within a given timing requirement or to ensure the robustness to missing, wrong or corrupted input measurements.

Table A.15. continued

Background and motivation

The introduction of SON functionalities is seen as one of the promising techniques for operators to save operational expenditures when deploying and operating new networks and in fact this concept is already being introduced in the context of E-UTRAN standardisation. It can also be envisaged that, with the increase of complexity of communication networks involving hundreds of parameters to be tuned, as well as with the introduction of high density of small sites (e.g. femto-cells), SON will become even more relevant.
SON concept has already received a lot of attention in recent years in 3GPP or in other initiatives such as the NGMN (Next Generation Mobile Networks) project. Recent works have formulated operator use cases for different stages such as planning (e.g. transmit power set-up), deployment (e.g. transport parameter set-up), optimization of the different radio parameters and maintenance (e.g. software upgrade), as well as for different techniques such as load balancing, interference coordination, packet scheduling or cell outage minimisation.
Targeting a fully SON as an ultimate objective, an evolutionary process can be envisaged, with the progressive inclusion of the SON culture. Given that changing the network configuration/parameterisation is a critical action from the operator's perspective, automated optimisation procedures are likely to be introduced in a step-by-step approach, starting with automatic sub-optimal behaviour detection mechanisms and establishing the corrective actions to overcome them through semi-automated optimisation of the different parameters, perhaps with some human intervention. An ultimate view, where a joint self-optimisation of several targets is performed will still require a lot of research efforts, due to the complex interactions between these targets. Supporting tools and/or models for this stage will also be needed to estimate the impact of parameter changes before making changes on the live network.

References

[1] Prehofer C., Bettstetter C.: Self-organization in communication networks: Principles and design paradigms. IEEE Commun. Mag. **43**(7), pp. 78–85 (2005).
[2] Bogenfeld E., Gaspard I.: Self-x in radio access networks. White Paper of the E3 project (2008).
[3] Hu H., Zheng X., Yang Y., Wu P.: Self-configuration and self-optimization for LTE Networks. IEEE Commun. Mag. **48**(2), pp. 94–100 (2010).

Table A.16. MP16. Cognitive and opportunistic positioning

Title MP16. Cognitive and opportunistic positioning

Author Sinan Gezici and Davide Dardari

Description

Cognitive and opportunistic positioning systems are characterized by spectrum aware capabilities in such a way that smart dynamic spectrum utilization is achieved (cognitive) and interfering signals present in the environment are suitably exploited to improve the positioning performance (opportunistic).
These systems provide various functionalities, such as seamless positioning, mobility management, location sensing, object detection and tracking, cognitive radar, and environment mapping [1].

Table A.16. continued

Background and motivation

So far the cognitive radio (CR) and opportunistic paradigms have been applied to the optimization of communication systems performance by adding awareness and responsiveness to interference and propagation effects. Furthermore, the availability of devices' position knowledge can be used to improve the performance of CR algorithms.
In cognitive and opportunistic positioning systems the same CR concept can be used to design positioning schemes with spectrum aware capabilities in such a way that smart dynamic spectrum utilization is achieved (cognitive) and interfering signals present in the environment can be suitably exploited to improve the positioning performance (opportunistic).
Therefore it is fundamental to design ranging/positioning algorithms that are able to adapt the ranging signal to the presence of the interference in order to optimize the performance by achieving dynamic spectrum utilization and adaptive transmissions. Some preliminary results show that the detection and avoid (DAA) approach does not always lead to the best performance but it may be preferable to allow a certain degree of spectrum overlapping between the interference and the useful signal [2].
The methodologies necessary to tackle this issue are:

- determine the theoretical bounds of ranging in the presence of interference or by positively exploiting the interference;
- given a certain interference measurement (from the CR sensing phase), determine how the ranging signal shape can be optimized in order to maximize the localization performance under power emission mask constraints;
- design practical low complexity cognitive ranging/positioning schemes;
- design practical low complexity opportunistic ranging/positioning scheme by exploiting the availability of signals coming from other systems (e.g., primary users).

References

[1] Celebi H., Arslan H.: Cognitive positioning systems. IEEE Trans. Wireless Commun. **6**(12), pp. 4475–4483 (2007).
[2] Dardari D., Karisan Y., Gezici S., D'Amico A.A., Mengali U.: Performance limits on ranging with cognitive radio. In: Proc. IEEE Int. Conf. Commun. (ICC), Dresden, Germany (2009).

Table A.17. MP17. Bringing dynamic spectrum access based on cognitive radio to reality

Title MP17. Bringing dynamic spectrum access based on cognitive radio to reality

Author Jordi Pérez-Romero

Description

Driven by the spectrum scarcity and at the same time the spectrum underutilisation identified in different frequency bands, one of the hot research trends in the last decade is to increase the efficiency of how the spectrum is being used. The so-called Dynamic Spectrum Access Networks (DSANs) consist of unlicensed radios, denoted as Secondary Users (SUs) who are allowed to operate in licensed bands provided that no harmful interference is caused to the licensees. The term Cognitive Radio (CR), originally coined in the late 90s, envisaged a radio aware of its operational environment and accordingly able to dynamically and autonomously adjust its radio operating parameters to adapt to the different situations. As a result, DSANs have been identified as a potential

Table A.17. continued

application of CRs, thanks to the ability to identify spatial and temporal spectrum gaps not occupied by primary users, and to place secondary/unlicensed transmissions within such spaces. Although a lot of fundamental research has been carried out, bringing DSA to reality will still require additional steps, involving not only technical aspects, but also significant regulatory changes.

Background and motivation

More than a decade has passed since the CR concept was introduced, and during this time a lot of research all over the world has been devoted to address different technical challenges of CR Networks applied to dynamic spectrum access, mainly covering fundamental problems. In that respect, different techniques have been developed, associated with sensing the presence of primary users, with deciding the appropriate frequency bands for secondary transmission and with enabling the proper handover mechanisms to vacate a channel when a primary user appears. The possibility of making use of databases to support the different cognitive radio procedures, as well as the availability of the necessary signalling mechanisms has also been studied. In the context of the utilisation of TV white spaces, IEEE 802.22 standard was developed including physical and medium access control layer for use by license-exempt devices on a non-interfering basis in spectrum portions allocated to the TV broadcast services. Although very relevant, the utilisation of TV white spaces reflects only a partial view of the problem since in this case the primary user is characterised by being static and by having a continuous transmission over a specific area. On the contrary, when trying to generalise the problem to other bands and to other types of primary users, there are still a lot of aspects that need to be solved such as how to cope with mobility, with the dynamicity in the traffic generation of the primaries, how to ensure specific QoS levels to secondary communications and how to coordinate the access of independent secondary systems.

References

[1] Mitola J. III, Maguire G.: Cognitive radio: Making software radios more personal. IEEE Personal Commun. **6**(4), pp. 13–18 (1999).
[2] Akyildiz I.F., Lee W.-Y., Vuran M.C., Mohanty S.: Next generation/dynamic spectrum access/cognitive radio wireless networks: A survey. Computer Networks, **50**(13), pp. 2127–2159 (2006).
[3] Sahai A., Cabric D.: A tutorial on spectrum sensing: Fundamental limits and practical challenges. In: Proc. IEEE Symp. New Frontiers in Dynamic Spectrum Access Networks (DySPAN), Baltimore, MD (2005).

Table A.18. MP18. Cooperation in wireless networks: Challenges and vision

Title MP18. Cooperation in wireless networks: Challenges and vision

Author Shlomo Shamai (Shitz)

Description

Cooperation in wireless networks defines a new concept which facilitates the enhancement in performance measured in terms of reliable communication rates, reliability, availability, as well as quality of service and coverage. Theoretical aspects of cooperation are described in [1], and the vision suggested focuses on those features, which are yet unsolved and are essential for the understanding of the practical potential and applications of these techniques.

Table A.18. continued

Background and motivation

Although the underlying theoretic concepts of wireless networks are well understood, cooperative systems are still in their infancy, and even more so is the unifying view of wireless networks. Fundamental research efforts are required as to understand the theoretical and practical promise and limitation of cooperation in wireless systems. Further, the theoretical understanding should be dramatically deepened in order to come up with efficient algorithms that are expected to come close to the theoretical limits, which at this point are not always fully characterized, and require further research.
In short we detail some fundamental problems that call for further basic research and understanding:

- *Random Channel State Information and the Impact of Channel Uncertainty.* Wireless systems in general and cellular communications in particular are modeled by random channel parameters, as for example fading, as well as dispersion in time and frequency. The usual assumption is that the channel is perfectly known at the receiving sites. While it is a good approximation, does not always reflect reality. Future work must consider the impact of channel uncertainty, and the cost of measuring the channels in the network. Channel measurement quality may be a factor in the degree and/span of cooperation and impact the optimal size of cooperating clusters. Future research on this aspect, should not come as an add-on component but be inherent with the wireless concept itself.
- *The Impact of Constrained Cooperation Links.* Of primary interest are wireless cooperation links, where cooperation is facilitated by backhaul connections to a central processing unit. While some theoretical results are available and were also accomplished as part of the NEWCOM++ efforts, fundamental research addressing limited backhaul bandwidth is an important area yet to be explored. Information-theoretic models provide tractable, elegant capacity formulas that are amenable to optimization, and performance bounds against which practical schemes can be compared. More importantly, however, they provide insights into the key performance bottlenecks, which can then be addressed in more practically oriented research.
- *Optimization of Wireless Cooperative Communications Networks under Practical Constraints.* Current theory and also practice account for elementary basic constraints, such as average power. In order to have a deeper theoretical guidance as for practical approaches, the theory should address also more practical constraints, starting with basic issue as per cell (base) or per antenna in a (multiple-Input-Multiple-Output) MIMO network constraints, and going through constrained bandwidth demands as well as peak amplitude, and the like. To the end of enhancing the insight to practical designs, joint optimization across many wireless terminals or hubs is required as well as cross-layer designs. In fact, full optimization should include all network/physical layer central aspects, as well as basic issues as synchronization and channel acquisition/tracing. Further, future challenges include coordinated power control, and the multi-terminal joint problems of scheduling, power control, and rate allocation across the frequency spectrum. Such procedures may require using state of art optimization theory, as well as practical routines. Further, many of these problems are computationally intractable, in general, and the way forward may be to look for structure of wireless networks that gives rise to duality concepts (such as multiple access (uplink) versus broadcast (downlink) duality concepts in MIMO networks). These and other methods facilitate theoretically simple and computable solutions.

References

[1] Kramer G., Marić I., Yates R.D.: Cooperative Communications. Foundations and Trends in Networking **1**(3-4), pp. 271–425 (2006).

Table A.19. MP19. Auction mechanisms for dynamic spectrum markets

Title MP19. Auction mechanisms for dynamic spectrum markets

Author Iordanis Koutsopoulos and George Iosifidis

Description

Recently, the liberalization of spectrum market was proposed as a solution for the problem of spectrum scarcity and the increasing unsatisfied demand of users for wireless services. In the near future it is expected that spectrum will be sold, leased or even exchanged very often between many different classes of entities. For example, spectrum regulators will sell their spectrum to operators which in turn, apart from providing services to their users, will be able to resale or exchange it with other operators.

Auction-based mechanisms will play a crucial role for the realization of these Dynamic Spectrum Markets. Auctions have been traditionally used for spectrum allocation by organizations such as the FCC in US. However, the emerging spectrum markets differ significantly due to the heterogeneity of the involved entities, the requirement for near-real time transactions and the spectrum reallocation option. Therefore, it is of high importance to consider these characteristics and devise appropriate auction algorithms for the allocation and the redistribution of the spectrum in order to satisfy the dynamic and diverse needs of users in future communication networks.

Background and motivation

Auction theory has been used for addressing many resource allocation problems in networks [1]. What makes auctions attractive is that they are agnostic to utility functions of the agents. Therefore, in the context of networks, the resource controllers (e.g. a server) are able through auctions to maximize the overall efficiency, i.e. the sum of nodes utilities, with minimum information requirement. Moreover, auctions can be used to increase the profit of the seller or to ensure certain resource allocation equilibria [2].

Spectrum auctions and especially auctions in the envisioned dynamic spectrum markets differ from traditional auctions of other goodies. In these setting, the auction designer should consider the spatial dimension and the fact that there may be many winners to which spectrum should be concurrently allocated. Spectrum reuse gives rise to *radio interference* due to mutual coupling of co-channel transmissions, and this in turn restricts the set of feasible channel allocations. Second, spectrum bands differ in terms of quality. Besides channel reuse, spectrum bands have different quality due to inherent frequency selectivity of the wireless channel, but also due to time-varying link quality and fading. Third, heterogeneity and unpredictability of user demands and mobility place additional challenges. Fourth, the small-scale, online fashion in which spectrum allocation needs to operate renders the machinery of bidding, allocation and payment a challenging task. Finally, there exists a wide range of methods for the allocation, each with its own assets. For example, spectrum can be divided in variable size bundles. There may be one or many auctioneers who may even compete. Or, spectrum may be exchanged with double auctions.

References

[1] Easley D., Kleinberg J.: Networks, Crowds, and Markets: Reasoning About a Highly Connected World. Cambridge University Press (2010)

[2] Huang J., Han Z., Chiang M., Poor H.V.: Auction-Based Resource Allocation for Cooperative Communications. IEEE J. Sel. Areas. Commun. (2008)

Table A.20. MP20. Capacity of multi-hop multi-flow networks

Title MP20. Capacity of multi-hop multi-flow networks

Author Giuseppe Caire

Description

Consider a discrete memoryless network with S sources, D destinations and N intermediate "relay" nodes, such that there exist some destinations $d \in D$ such that their observation Y_d is not independent of the relay node inputs $\{X_n : n \in N\}$, given the source inputs $\{X_s : s \in S\}$. In general, such network is called "multi-hop" since the sources do not control completely the destination signals. In addition, we require that each destination d has a list of desired sources S_d, that define the probability of error at d: the destination d is in error if any decoded message from sources in S_d is in error. If the sets S_d are not identical for all destinations, the network is called "multi-flow". Such network is fundamentally characterized by the presence of interference: the messages of the sources s not in S_d act as interference with respect to destination d.
At this point, the problem can be stated as: find the capacity region of such multi-hop multi-flow network, in general or at least in some non-trivial special cases, either exactly or in some meaningful approximated sense. In particular, when the network is formed by linear-Gaussian channels defined over the real or the complex field, we are in the presence of a multi-hop multi-flow Gaussian network, useful to model wireless communication networks with relaying, cooperation and interference. In this case, an approximated characterization of the network capacity region requires to finds an achievable region that differs from some outer bound by some bounded constant, that may depend on the network topology but it is independent of the channel coefficients.

Background and motivation

Classical Information theory has successfully characterized the ultimate limits of reliable communications for single-source single-destination single-hop settings (i.e., the so-called point-to-point channel). Single-hop networks with multi-source (many senders) and one or more destinations that wish to decode all messages are also well characterized, resulting in the well-known multiple-access channel (MAC) and variants thereof. Single-hop networks where different receivers wish different subsets of the messages are generally still open (e.g., the general broadcast channel, the general interference channel, and the index coding problem in network coding). Yet, several important advances have been made in single-hop networks, as for example the recent discovery of the "interference alignment" principle. Finally, multi-hop networks with multiple sources and multiple destinations, where the destinations wish to decode all the messages (the so-called multi-source multi-cast problem) has also been recently characterized in the Gaussian case, with inner and outer bounds to the capacity region that differ by a fixed gap, independent of the channel coefficients, and in the noiseless network coding case, where it is known that linear random network coding achieves the min-cut max-flow bound. In contrast, the most general setting of a multi-hop network with multiple sources and destinations, where the destinations wish to decode different subsets of the source messages, remains widely open. In this case, only very preliminary results exist and no general theory and achievability scheme has yet been developed. Furthermore, some negative results exist: for example, it is known that even in the noiseless case of a network of perfect point-to-point links of fixed capacity, linear network coding is generally suboptimal. The general multi-hop multi-flow network, at present, remains the "holy grail" of multi-terminal information theory, and solutions (even in the approximate finite gap sense) of this problem are expected to provide deep fundamental insight on how to design network architectures grounded on fundamental principles rather than on heuristics, as mostly done so far.

Table A.20. continued

References

[1] Cadambe V.R., Jafar S.A.: Interference alignment and degrees of freedom of the K-user interference channel. IEEE Trans. Information Theory **54**(8), pp. 3425–3441 (2008).
[2] Avestimehr A.S., Diggavi S.N., Tse D.N.C.: Wireless network information flow: A deterministic approach. IEEE Trans. Information Theory **57**(4), pp. 1872–1905 (2011).
[3] Lim S.H., Kim Y.-H., El Gamal A., Chung S.-Y.: Noisy network coding. In: Proc. Information Theory Workshop (ITW), Dublin, Ireland (2010).

Table A.21. MP21. Structured coding in wireless networks

Title MP21. Structured coding in wireless networks

Author Shlomo Shamai (Shitz)

Description

As of late the potential of structured coding for multi-terminal wireless networks has been identified. This in fact deviates from the practice over the last decades, where coding methods were developed in an attempt to reach the limits of communications governed by the Shannon capacity, that, for many problems, as point-to-point communications, can be shown to be achieved by random coding arguments, which are formidable for any practical considerations. This dates back to Shannon's original 1948 work, and accounts for the very recent developments of turbo and LDPC coding approaches, which indeed come close to the capacity of these point-to-point links.

Background and motivation

For networks, that is multi-terminal communication schemes, including, multiple-access, broadcast, interference, relay channels, state dependent multi-terminal channels, and combinations of the above as specific examples, this is not the case. Apart from special cases, the information-theoretic limits of these systems are not fully known yet. Recently in a variety of cases, it has been shown that structuring, like lattice based codes, algebraic codes, and the like, can in fact provide serious theoretical (not only practical) advantages over classical random coding.
While beautiful examples have been constructed in a variety of applications, the research in this domain is at its infancy yet, and future research efforts are called for as to acquire the insights required towards creating a wide scope applied technology. Structuring will in fact apply to many aspects, and centrally play a role in interference mitigation, as is demonstrated by recent advances in the theoretical research of interference alignment in wireless systems. These ideas are also fundamental in network coding, and in fact provide a theoretical bridge between what has been considered different domains of research.
Challenging specific research aspect within this setting could focus on possible polarization (or even distributed polarization), which accounts within the process for the required structuring, and that is very different from our understanding at this level.

Table A.22. MP22. Coding and feedback

Title MP22. Coding and feedback

Author Erdal Arikan

Description

Code designers have traditionally focused on channel models without feedback. In particular, the capacity-achieving turbo and LDPC codes were originally developed for one-way channels. These codes have been adapted to systems with feedback by various H-ARQ techniques. This approach can be justified in systems where feedback is haphazard or suffers high latency. However, modern wireless system designs, notably LTE-A, have frame structures designed for low-latency feedback. It is time to develop new coding methods that incorporate feedback as an integral part of the code design, rather than an accessory.

Background and motivation

This problem is important because it has the potential of delivering more reliable communication at reduced complexity and lower latency. It may also provide higher throughput.

- *Complexity.* It is well known that feedback simplifies the coding problem; in fact, perfect feedback may eliminate it entirely. For example, a simple retransmission scheme can achieve channel capacity over an erasure channel provided there is timely reliable feedback.
- *Lower latency.* Error-correcting codes provide reliability at the expense of latency. If there is feedback, latency can be reduced by using shorter codes.
- *Higher throughput.* In many communication systems the transmitter does not know the channel state in advance. If there is no feedback, one has to be conservative in choosing the modulation and coding scheme so as to avoid transmitting at rates above channel capacity. If one knows that feedback about the fate of the transmission will soon become available one may transmit at high rates initially so as to take advantage of the full capacity that the channel may offer momentarily.

Code design for a time-varying channel with feedback may combine a number of existing techniques from information theory and coding practice. Among the tools that we foresee to be useful are combined source-channel coding (to predict the performance when transmitting at rates above capacity), turbo and LDPC code design and iterative decoding techniques.

References

Since we advocate going back to fundamentals for code design in the presence of feedback, we list here Shannon's classic paper on two-way channels and a more recent paper that gives the state-of-the-art with pointers to the literature.

[1] Shannon C.E.: Two-way communication channels. Proc. Fourth Berkeley Symp. on Math. Statist. and Prob. **1** (Univ. of Calif. Press, 1961), pp 611–644.
[2] Gallager R.G., Nakiboglu B.: Variations on a theme by Schalkwijk and Kailath. IEEE Trans. Information Theory **56**(1), pp. 6–17 (2010).

Table A.23. MP23. Networks with traffic accumulation

Title MP23. Networks with traffic accumulation

Author Daniel Costello, Martin Haenggi, and Joerg Kliewer

Description

Networked wireless communications over multiple hops is rapidly emerging as the main architecture of future wireless systems, including multihop extensions of cellular and WiFi networks, mesh networks, and sensor networks. Common among these types of networks is that they are not completely unstructured (or ad hoc) networks, but traffic is routed and accumulated towards a common destination. Due to this characteristic property, such networks can be referred to as Networks with Traffic Accumulation, or NETAs. Traffic accumulation creates hot spots or bottlenecks around the common destination because of the increased traffic load and interference. Despite the severity of the hot spot problem, no efficient strategies to cope with it have been proposed in the literature.

Background and motivation

While several aspects of multihop transmissions are now well understood, there are currently no efficient methods to mitigate this hot spot problem. One proposed solution is to address this issue by developing new distributed error correction strategies for NETAs. Rather than employ existing coding schemes for this challenging environment, we believe it will be necessary *to find and study new coding strategies that are tailored to this emerging class of networks.*
To the best of our knowledge, efficient coding solutions for NETAs with arbitrary topologies do not exist in the literature, although a special case has recently been considered in [1] for a fixed tree-type structure of depth three using parity check codes. In the case of sensor networks, researchers have studied the use of existing coding schemes and identified their shortcomings. In [2], for example, convolutional codes are analyzed for power efficiency in a frequency-flat, slow Rayleigh fading channel. It was found that the energy required to encode data is negligible. As a consequence, using forward error correction (FEC) is inefficient if Viterbi decoding is performed using a dedicated onboard microprocessor. The conclusion is that, if energy-efficiency is critical, convolutional codes may not be the best solution. In [3], block and convolutional codes are studied in three scenarios: (1) direct transmission from source to destination; (2) transmission via the intermediate nodes with FEC decoding only at the final destination; (3) transmission via the intermediate nodes with FEC decoding at each node. The optimum scheme is shown to depend on the inter-node distances and on the relative energy cost of decoding. [4] suggests the use of LDPC codes to reduce transmission power consumption. It shows that LDPC codes are 42% more energy-efficient than BCH codes (which in turn are more energy-efficient than convolutional codes). However, they are not concerned with the limited computational capabilities of (typical) sensor network nodes. [5] is a higher-level paper that promotes single-hop communication to a base station that is assumed to be wall-powered. It promotes an asymmetric design with a high-power uncoded downlink to the nodes and a low-power coded uplink to the base station that uses a code with low encoding complexity and high decoding complexity. [6] is one of the few (perhaps the only) journal article devoted to the subject of FEC in sensor networks. It studies the critical link distance at which FEC becomes power-efficient. The analysis essentially trades off the coding gain and the decoding complexity. It shows that in many situations FEC is only beneficial for distances that are not typical for sensor network applications, unless power amplifier efficiencies are factored in. Their analysis does not include interference, however. Their conjecture that interference would significantly reduce the critical distance is certainly in agreement with intuition. Since interference is quite likely the performance-limiting factor in NETAs, in particular in the area around the base station, interference must be included in the analysis of proposed coding strategies. [7] introduces a cross-layer (physical, MAC, networking) methodology for the analysis of error control schemes

Table A.23. continued

in wireless sensor networks and uses it to compare FEC using BCH codes and error detection with automatic repeat request (ARQ) for Mica2 and MicaZ Berkeley motes. If the nodes' positions are known and channel state information (CSI) is available at the transmitters, they show that a transmit power reduction (or longer hops) is possible if FEC is employed. The analysis is based on the expected, not the actual (random), hop distances, and the fact that, for this type of hardware, a transmit power reduction does not reduce total power consumption is not considered. In summary, prior research has revealed that the use of conventional FEC strategies is not necessarily beneficial for energy-constrained sensor nodes. We strongly believe, however, that significant coding gains are achievable with new coding strategies that are tailored to the specific properties of NETAs.

References

Since we advocate going back to fundamentals for code design in the presence of feedback, we list here Shannon's classic paper on two-way channels and a more recent paper that gives the state-of-the-art with pointers to the literature.

[1] Kim S.W.: Concatenated network coding for large scale multi-hop wireless networks. In: Proc. IEEE Wireless Commun. & Networking Conf., Hongkong, China (2007).

[2] Shih E., Cho S., Ickes N., Min R., Sinha A., Wang A., Chandrakasan A.: Physical layer driven protocol and algorithm design for energy-efficient wireless sensor networks. In: Proc. ACM Mobi- Com, Rome, Italy (2001).

[3] Zorzi M., Rao R.R.: Coding tradeoffs for reduced energy consumption in sensor networks. In: Proc. IEEE Personal, Indoor and Mobile Commun. (PIMRC), Barcelona, Spain (2004).

[4] Sartipi M., Fekri F.: Source and chanel coding in wireless sensor networks using LDPC codes. In: IEEE Conf. Sensor and Ad Hoc Communication and Networks (SECON), Santa Clara (2004).

[5] Zhong L., Rabaey J., Wolisz A.: Does proper coding make single hop wireless sensor networks reality: The power consumption perspective. In: Proc. IEEE Wireless Commun. & Networking Conf., New Orleans, LA (2005).

[6] Howard S.L., Schlegel C., Iniewski K.: Error control coding in low-power wireless sensor networks: When is ECC energy-efficient? EURASIP J. Wireless Comm. Networking, pp. 1–15 (2006).

[7] Vuran M., Akyildiz I.: Cross-Layer analysis of error control in wireless sensor networks. In: IEEE Conf. Sensor and Ad Hoc Communication and Networks (SECON), Reston, VA (2006).

Table A.24. MP24. Channel estimation for user aware networks

Title MP24. Channel estimation for user aware networks

Author Luis M. Correia

Description

One can envisage that future wireless networks will take advantage of users terminals as sensors to radio channels, and then use this information to estimate channels for the whole servicer area, hence, improving the quality of service in networks and quality of experience for users. This approach implies work on several areas: three-dimensional estimation of radio channels, real-time signal processing for three-dimensional radio channels reconstruction, cross-layer design for the usage of radio channel information in the optimisation of user data transfer, usage of radio channel information in the optimisation of radio resource management.

Table A.24. continued

Background and motivation

Given that the increase of data rate will still be an important parameter in the evolution of wireless systems, and that spectral efficiency theoretical limits is being reached, other dimensions and techniques need to be addressed. The spatial dimension has already been addressed, like antenna beamforming, Spatial Division Multiple Access, and MIMO, but still another perspective has to be studied: three-dimensional channel estimation. This way, networks can take advantage of information obtained via multiple users (taken as channel sensors) so that a "full view" of the channel can be obtained, and then, used to optimise the link between each user and a base station.

Table A.25. MP25. Fundamental Issues in Information Theory.

Title MP25. Fundamental Issues in Information Theory

Author Sergio Verdú

Description

Entropy rate of sources with memory: Other than Markov sources, the entropy rate of most discrete sources with memory is unknown. For example, the entropy rate of a binary symmetric Markov chain with transition probability ρ seen through a binary symmetric channel with crossover probability δ (i.e., a hidden Markov chain) is not known as a function of ρ and δ.
Interference channels: The capacity region of the interference channel with additive independent Gaussian noise is unknown.
Rate-distortion function of sources with memory: Not even for a simple binary symmetric Markov chain with bit error rate criterion, the rate-distortion function is known.
Capacity of the deletion channel: The simplest channel whose capacity is unknown is a binary channel with memory where each input bit is either reproduced perfectly at the output or deleted with probability δ, independently of what happened to other bits. If the locations of the deleted bits are known at the decoder, the channel is the binary erasure channel whose capacity is $1-\delta$. Otherwise, the capacity is unknown.
Impact of feedback on channel capacity: Shannon showed in 1956 that feedback does not increase the capacity of memoryless channels. If the channel has non-white Gaussian noise, the capacity increase due to feedback is unknown except in simple cases

References

[1] Blackwell D.: "The entropy of functions of finite-state markov chains." Trans. Prague Conf. Information Theory, Statistical Decision Functions, Random Processes, pp. 13–20 (1957).
[2] El Gamal A., Kim Y.H.: Network Information Theory. Cambridge Univ. Press, Cambridge, UK (2011).
[3] Gray R.M.: "Rate distortion functions for finite-state finite-alphabet Markov sources." IEEE Trans. Information Theory **17**(2), pp. 127–134, (1971).
[4] Mitzenmacher M.: "A survey of results for deletion channels and related synchronization channels." Probability Surveys, pp. 1–33 (2009).
[5] Kim Y.H.: "Feedback capacity of the first-order moving average Gaussian channel." IEEE Trans. Information Theory **52**(7), pp. 3063–3079 (2006).

Acronyms

ACGN	Additive Coloured Gaussian Noise
ADC	Analogue to Digital Converter
AMC	Adaptive Modulation and Coding
APP	A-Posteriori Probability
ASIC	Application Specific Integrated Circuits
ASIP	Application Specific Instruction-set Processors
AWGN	Additive White Gaussian Noise
BAN	Body Area Network
BCJR	Bahl-Cocke-Jelinek-Raviv
CDMA	Code Division Multiple Access
CPC	Cognitive Pilot Channel
CR(N)	Cognitive Radio (Network)
CSI	Channel State Information
CSMA	Carrier Sensing Multiple Access
(M)CRB	Modified Cramér-Rao Bound
DMC	Discrete Memoryless Channel
DSA	Dynamic Spectrum Access
DTN	Delay Tolerant Network
DSP	Digital Signal Processor
DVB-T,S,H	Digital Video Broadcasting-Terrestrial, Satellite, Handheld
FSM	Flexible Spectrum Management
FC	Fusion Centre
FEC	Forward Error Correcting
FFT	Fast Fourier Transform
FTM	Filtered MultiTone
GALS	Globally Asynchronous Locally Synchronous
GC	Green Communications
GMI	Generalized Mutual Information
PP	General Purpose Processor
IBP	Iterative Belief-Propagation
IC	Integrated Circuit

ICI	Inter-Carrier Interference
ICT	Information Communication Technology
IDMA	Interleave-Division Multiple Access
IQ	Input-Queued
ITU	International Telecomunications Union
JSCC/D	Joint Source and Channel Coding/Decoding
LDPC	Low Density Parity-check Codes
KPI	Key Performance Indicator
LTE	Long Term Evolution
LUT	Look-Up Table
MAC	Medium Access Control
MANET	Mobile Ad-hoc NETwork
MAP	Maximum A-Posteriori
MC	MultiCarrier
MCP	Multi-Cell Processing
MIMO	Multiple Input Multiple Output
MUD	MultiUser Diversity
NBS	Nash-Bargaining Solution
NC	Network Coding
NE	Nash Equilibrium
NOFDM	Non-Orthogonal Frequency Division Multiplexing
OBU	On-Board Units
OFDM	Orthogonal Frequency Division Multiplexing
OFDMA	Orthogonal Frequency Division Multiple Access
OODA	Observe, Orient, Decide and Act
PAPR	Peak-to-Average Power Ratio
PIM	Platform Independent Model
PPP	Poisson Point Process
PU	Primary Users
QAM	Quadrature Amplitude Modulation
QoE	Quality of Experience
QoS	Quality of Service
RAT	Radio Access Technology
RF	Radio Frequency
RFID	Radio Frequency IDentification
RRM	Radio Resource Management
RRS	Reconfigurable Radio Systems
RSU	RoadSide Units
RX	Receiver
SCA	Software Communications Architecture
SDR	Software Defined Radio
SISO	Single-Input-Single-Output
SIC	Successive Interference Cancellation
SNR	Signal-to-Noise Ratio
SoC	System on Chip

SON	Self Organising Network
SOVA	Soft-Output Viterbi Algorithm
SU	Secondary User
TOA	Time Of Arrival
TX	Transmitter
UWB	Ultra Wide Band
VANET	Vehicular Ad hoc NETwork
VoD	Video On Demand
VoIP	Voice Over IP
VOQ	Virtual Output Queueing
WAVE	Wireless Access in Vehicular Environments
WLAN	Wireless Local-Area Networks
WSN	Wireless Sensor Network
WWRF	Wireless World Research Forum

MIX
Papier aus verantwortungsvollen Quellen
Paper from responsible sources
FSC® C105338

If you have any concerns about our products,
you can contact us on
ProductSafety@springernature.com

In case Publisher is established outside the EU,
the EU authorized representative is:
Springer Nature Customer Service Center GmbH
Europaplatz 3, 69115 Heidelberg, Germany

Printed by Libri Plureos GmbH
in Hamburg, Germany